Surveillance and Reconnaissance Imaging Systems

Modeling and Performance Prediction

For a list of recent titles in the *Artech House Optoelectronics Library*, turn to the back of this book.

Surveillance and Reconnaissance Imaging Systems

Modeling and Performance Prediction

Jon C. Leachtenauer
Ronald G. Driggers

Artech House
Boston • London
www.artechhouse.com

Library of Congress Cataloging-in-Publication Data
Leachtenauer, Jon C.
    Surveillance and reconnaissance imaging systems : modeling and performance
prediction / Jon C. Leachtenauer, Ronald G. Driggers.
        p.   cm. — (Artech House optoelectronics library)
    Includes bibliographical references and index.
    ISBN 1-58053-132-6 (alk. paper)
    1. Electronic surveillance.   2. Imaging systems.   3. Space surveillance.
  4. Aerial reconnaissance.   I. Driggers, Ronald G.   II. Title.   III. Series.
    TK7882.E2 L43   2001   00-068932
    621.389'28—dc2l   CIP

British Library Cataloguing in Publication Data
Leachtenauer, Jon C.
    Surveillance and reconnaissance imaging systems : modeling and performance
    prediction
    1. Electronic surveillance   2. Imaging systems
    I. Title   II. Driggers, Ronald G.
    621.3'8928

    ISBN 1-58053-132-6

Cover design by Gary Ragaglia

© 2001 ARTECH HOUSE, INC.
685 Canton Street
Norwood, MA 02062

International Standard Book Number: 1-58053-132-6
Library of Congress Catalog Card Number: 00-068932

10 9 8 7 6 5 4 3 2 1

To Mary Ellen, Caroline,
Amy, Jon, and Paul

Contents

Preface

This text is intended to provide a comprehensive review of surveillance and reconnaissance (S&R) imaging system modeling and performance prediction. The goal of S&R systems is to provide information—detection, classification, and identification of objects and features based on image characteristics. Given a set of information requirements, the system designer and operator must design and operate a system in a manner that will provide the required information. They thus require the ability to model and predict the performance of S&R systems based on design attributes and measurable operating parameters. Performance is defined in terms of the ability of users to extract the desired information. System descriptions, S&R modeling history, system characteristics, and performance models are all provided. While there are many texts describing imaging systems in the realm of linear shift invariant systems, target acquisition, driving, and flying, there are few texts that address the specific design and analysis techniques used with S&R imaging systems. The emphasis in this book is on validated prediction of human observer performance.

This book is intended for the design engineer, system analyst, and image scientist, and may also be useful to the manager or project engineer who is interested in gaining a broader understanding of image system modeling and learning the nomenclature. Human factors practitioners involved in observer performance measurement may also find the book of interest.

The book begins with a description of the conceptual performance prediction model and a review of alternative modeling approaches. A brief description of S&R systems that includes the visible, infrared, and synthetic aperture radar (SAR) imagers is provided in Chapter 2. A historical review is provided in Chapter 3, which summarizes the evolution of both S&R systems and S&R performance prediction from WWI to the present. Performance prediction approaches begin with the simplistic scale and resolution rules-of-thumb to today's information theory and multivariate spatial and frequency domain models. Linear shift-invariant (LSI) systems are covered in Chapter 4 and include both the spatial domain and frequency domain representations.

The variables used to measure S&R information extraction are provided in Chapter 5 in terms of observer performance or dependent variables. The dependent variables include both direct measures of information extraction performance and predictions of performance using subjective scales. Measures such as the probabilities of task discrimination (detection, recognition, and identification) and detection time and range metrics are direct measures of information extraction. The role of signal detection theory in modeling the tradeoff between detection and false alarms is briefly reviewed. An alternative to the direct measurement of performance is the use of predictive measures such as the National Imagery Interpretability Rating Scale (NIIRS). These measures represent performance predictions by trained observers. Classical information theory has also been applied to the problem, generally as an intermediate step to predicting information extraction performance. The predictor (independent) variables are reviewed in Chapter 6. These independent predictor variables correspond to quantities like resolution, sharpness, and scale, and are both measurable image properties and predictable from system design and operating parameters. Other independent variables discussed include contrast, signal-to-noise, and difference and quality attribute metrics.

The performance of an S&R system, although largely a function of its design and operation, is also affected by other elements of the imaging chain. Target and environmental considerations are described in Chapter 7. In addition to these effects, some basic considerations are given to the modeling of deception and denial techniques. These methods include camouflage, dummies, and signal reduction techniques such as aerosols. Chapters 8 and 9 provide descriptions and models for image processing and display/observer considerations, respectively.

In Chapter 10, documented models used to predict the performance of S&R systems are reviewed. Models are provided for SAR, electro-optical/visible and infrared systems, and video systems. In Chapter 11, conversion methods between alternative modeling approaches are discussed as a means to perform system comparisons. Finally, unresolved issues with modeling the performance of S&R systems are described in Chapter 12. These include modeling of multi- and hyper-spectral system performance, SAR observer performance, treatment of temporal effects, and modeling of search.

Acknowledgements

As I embarked on a long and rewarding career, I had the good fortune to receive an introduction to the human factors profession from Dr. Kenneth Cook, Dr. Robert Sleight, and Mr. Henry Abbott of Century Research Corporation. I am indebted to their patient nurturing of my interests. My education continued with Dr. Sheldon McCloud of the Rome Air Development Center. At Boeing, I was fortunate to enjoy continued interaction with a strong group of professionals including Dr. Jim Briggs, Dr. Charles Elworth, Dr. Conrad Kraft, Mr. John Booth, Mr. Richard Schindler, and Mr. Richard Farrell. At what is now Veridian ERIM-International, I enjoyed the support of and continued my education with Dr. Ralph Mitchel, Dr. Jack Walker, Mr. Basil Wentworth, Mr. Rod Dallaire, and Ms. Linda Klimach. I have learned much in my work with the National Information Laboratory from Mr. Michael Grote, Mr. Michael Brill, Dr. Ron Enstrom, Mr. Dennis Bechis, Mr. Albert Pica, and Dr. Jeff Lubin. At the National Exploitation Laboratory, and subsequently the National Imagery and Mapping Agency, I was fortunate to enjoy a long association with Mr. Michael Parcell. I am grateful to all of the many people from whom I learned so much over the last 30 years including the many that I have failed to mention.

The idea for this book was originally suggested by my co-author, Dr. Ron Driggers as a result of some earlier collaborations. I appreciate Ron's patience in guiding me through the effort. Finally, I would like to mention the very valuable editing help provided by Ms. Ellen Schwartz and Mr. Keith Krapels.

1

Introduction

Surveillance and reconnaissance (S&R) systems are defined here as remote sensing imaging systems used to acquire military, economic, and political intelligence information. Classically, reconnaissance is defined as the act of reconnoitering or making a preliminary inspection. In the military sense, it involves determining the lay of the land and the disposition of enemy forces. Economically, it may be a survey to detect oil-bearing strata. Any imaging system that can acquire imagery of relatively large ground areas can be used for reconnaissance.

Surveillance is defined as maintaining close observation of a group or location. Frequent imaging of enemy forces is the classic application. Monitoring crop vigor or civil unrest can also be considered surveillance. The implication here is the need for frequent or even continuous coverage.

In a practical sense, most S&R systems can perform both reconnaissance and surveillance by virtue of the ability to trade off resolution and area of coverage. The Predator unmanned aerial vehicle (UAV), for example, flies a video system with both a long focal length lens for high resolution surveillance and a short focal length lens for lower resolution reconnaissance. The LANDSAT multispectral satellite is used for both reconnaissance and surveillance applications, the only difference being the number of images acquired of the same ground area.

Whether the mission is reconnaissance or surveillance, the goal is to image an area or object on the ground and to extract information from that image. Relating the ability to extract information to the design and performance of imaging systems is the topic of this book.

S&R imaging systems range from synthetic aperture radar (SAR) to infrared (IR) and visible systems. In the IR and visible portion of the spectrum, they include video systems as well as the conventional line scan and push-broom sensors. The focus is on electro-optical (EO), as opposed to photographic systems, although some of the approaches are applicable to both. The vast majority of models and performance predictors currently available deals with monochrome imagery. Although multi- and hyperspectral systems are currently flown, the spectral contribution of such systems has not been well defined or modeled.

The performance predictions discussed in this book assume a human observer as the extractor of information from the image. Despite some 30 years of research on machine or automated image exploitation, there are no fully functional automated systems yet available.

Performance prediction is based on some type of modeling process. At its simplest, this process may be a simple regression model relating some measure of image quality to a measure of information extraction performance. At the complex end of the spectrum, the model may include computer simulations of the imaging process or theoretical models of the human visual system.

While this book treats a wide variety of performance prediction approaches, it focuses on two. They are the general image quality equation [1], developed for generally higher altitude EO systems, and the probability of discrimination model [2], developed for ground and low altitude EO systems. A variety of other approaches are also presented in sufficient detail to allow the reader to assess their applicability to particular performance prediction problems.

1.1 Modeling Applications

There are two reasons for predicting the performance of an S&R system. The first is as a tool in the system design process. Given a set of information requirements, one would like to design a system that could meet just those requirements. In a competitive process, overdesign is as costly as underdesign. In the sensor design process, the whole chain from target or scene to human information extraction needs to be well understood (and modeled to varying degrees) so as to optimize the sensor system design. The sensor does not operate independently, but is part of an overall system.

A second reason for predicting performance is in the operation of an S&R system. Most systems have some flexibility in their deployment and operation. Aircraft systems, for example, can fly at different altitudes and thus trade area coverage and resolution. In a hostile environment, the pilot would generally like to fly as high as possible while still acquiring imagery of sufficient quality to satisfy the information requirement. Image scale, which is the ratio of image to ground distance, was one of the first image quality metrics used to predict system performance. Given the focal length of the camera, required imaging altitude could be calculated.[1] Target range (target-to-sensor distance) is a common metric for tactical imaging systems.

As performance prediction has improved, it has become more complex. Atmospheric effects can be modeled and used in predicting the performance of operational systems. Target/sensor geometry effects also enter the prediction equation. Performance losses occurring in the transmission and display of images can be estimated and used as part of the system tasking process.

Although somewhat outside the scope of this book, performance prediction for the purpose of understanding and optimizing imaging system performance is a useful tool. Relatively simple physical and computer models are used to show the effects of changes in system operation (e.g., effects of sun angle on shadows, target layover for SAR, etc.). Such models lead to a better understanding of the imaging process and thus improve information extraction; they are also used to optimize system tasking for a particular application.

1.2 Conceptual S&R Model

The conceptual model of the S&R system predicts some type of information extraction performance based on results of the imaging operation (Figure 1.1). Performance is defined using one or more measures of information extraction. These measures are discussed in detail in Chapter 5. Information extraction can be measured directly by exposing observers to imagery and recording their information extraction performance (e.g., number of targets detected). We will use the terms "direct performance measurement" and "direct performance measures" to refer to this method of measurement. Direct performance measures include such things as the probability of detection and identification. They are termed direct because they are measured by scoring actual performance. A target is detected or it is not. A target is correctly identified or it is misidentified. Observers can also provide information extraction

---

1. Image scale is generally defined as a ratio (e.g., 1/18,000). The numerator is termed the scale factor. Scale factor for a camera is defined as altitude divided by focal length (both in the same units). Therefore, focal length times the scale factor defines the required altitude.

performance estimates using various rating scales. We will use the term "performance estimate measurement and measures" to refer to this category of data and method of measurement. The terms "objective" and "subjective" have often been used to refer to direct measures and estimated measures, respectively. Direct measures are considered by some to be objective because responses are either right or wrong and can be directly scored as such. Performance estimates are considered subjective because they represent the opinion of the observer (e.g., "I believe this task can be performed"). Although the direct measurement approach seems more rigorous, in actual practice such measures may be less accurate or precise (in terms of predicting operational performance) than performance estimates. Although direct measurement responses can be scored objectively, the underlying response can be quite subjective. The theory of signal detection literature shows how direct performance can vary solely on the basis of an observer's decision criteria.

Performance estimate measures range from performance predictions to statements of relative performance or goodness. Performance predictions include probability of performance estimates (estimates of the probability of detection in imagery of a defined quality level), task satisfaction confidence ratings, and National Imagery Interpretability Rating Scale (NIIRS) ratings. Task satisfaction confidence ratings indicate the observer's confidence that a given task can be performed on the image viewed (or on one of like quality). The NIIRS [3] is a 10-level scale with each level defined by a series of increasingly difficult information extraction tasks (e.g., detect tanks, identify

Imaging operation + Information extraction = Information

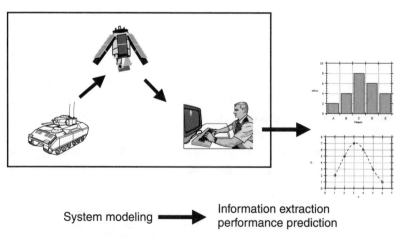

System modeling ➡ Information extraction performance prediction

Figure 1.1 Conceptual S&R model.

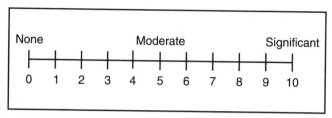

Figure 1.2 Relative quality scale.

small helicopters). A rating of a given value indicates that the tasks at the defined levels, as well as all those lower (and easier) tasks, can be performed. Finally, relative performance can be defined in terms of verbally anchored scales. Figure 1.2 provides a typical example. The scale is a 10-point scale with three verbal anchors.

In order to model or predict performance, one or more predictor variables are needed. These variables relate to the spatial and tonal characteristics of imagery (resolution and contrast, for example). They may be based on image measurements (physical image quality) or on image differences (pixel amplitude differences). They may also be based on modeling information in the sense of mathematical or physical models.

The conceptual model of the S&R system considers what is called the image chain. The image chain (Figure 1.3) is a sequence of actions and phenomena affecting the information that can be extracted from the results of

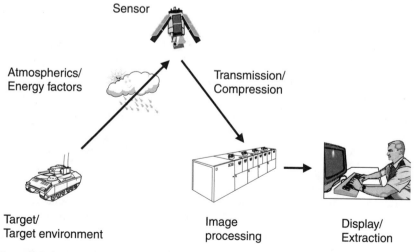

Figure 1.3 Image chain.

an imaging operation. The chain begins with a scene (target and environment) and progresses to the perceptual and cognitive task of extracting information. Referring to Figure 1.3, the chain begins with objects of interest in a scene. Objects of interest are referred to as targets. Target characteristics (size, shape, spectral characteristics) all affect the information that can be extracted from a given image. Performance is also affected by the uniqueness of a target relative to both its background (target conspicuity) and to other similar targets. In identifying automobiles, for example, the Volkswagen Beetle is more unique than a Toyota Camry. Similarly, a bright red car tends to be more conspicuous, and more easily detected, than a gray car.

The characteristics of the target environment affect target contrast or target/background separation. A target under tree canopy or camouflage is more difficult to detect and identify than one in the open. A homogeneous background (e.g., pavement) is generally more helpful in target detection than a cluttered background (e.g., scrub vegetation).

The next set of factors to be considered relate to the imaging process exclusive of the sensor itself. This set would include atmospheric effects, imaging geometry effects, and what we term "energy-related factors." Atmospheric effects include energy attenuation due to transmission losses, blur due to turbulence and aerosol scattering, and distortion related to the atmosphere. Haze is considered an atmospheric effect. Geometry effects include distance and angle. For a given IR or visible imaging system, increasing target-to-imager distance clearly degrades information extraction performance. This is generally not the case for coherent systems such as SAR. The angle of a target relative to the imaging plane defines those components of the target that can be seen (and imaged) and can also affect the size of the target image. A top view of an aircraft will normally be larger than a side view. A top view also tends to be more distinctive (Figure 1.4). The term "energy-related" is used here to describe those factors other than the atmosphere that can affect the target/background energy relationship and thus contrast. For a visible imaging system, this includes sun angle effects. For an IR system, factors affecting thermal contrast such as weather history and wind effects have an impact.

The actual imaging operation records the target/background energy relationship as modified by the atmosphere and environment. The operation entails some type of energy transform (light or thermal energy) to electrical signals, for example. A viewable image may be formed at the sensor, or energy may simply be captured and transmitted for further processing. The characteristics of the imaging sensor, of course, play a major part in system performance. The ability of the sensor to capture fine spatial and spectral

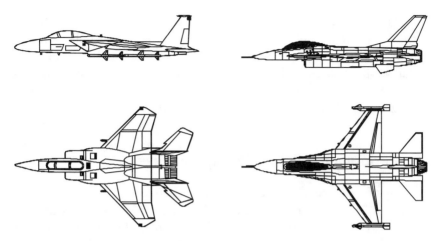

Figure 1.4 Effect of aspect angle.

detail with minimal noise is one of the most critical aspects of the image chain.

If data captured by the sensor are transmitted to the ground, the transmission process can affect final performance. Because of limited bandwidth, it is common to apply some type of bandwidth compression to the transmitted data. To the extent that compression is not lossless[2] (i.e., is lossy), performance degradation results.

Initial ground processing generally provides a viewable image, either in hardcopy (film) or softcopy (cathode ray tube or flat panel display). It is at this stage that physical image quality measurements can be made in an attempt to predict information extraction performance. Although it is true that the initial image may be transmitted and redisplayed multiple times, this operation will not be discussed further.

The viewable image must be displayed and information extracted from that image. The display process itself affects performance and includes both the display and the display environment. An image may be displayed on a high-quality CRT, but if the CRT is in a brightly lit room, contrast is decreased and information extraction performance degraded.

---

2. Lossless compression is defined as a compression where the reconstructed image is not perceptibly different than the original image and, thus, no information is lost. Differences are typically measured using some type of subjective comparison or rating scale. The two images may differ in terms of absolute values, but so long as those differences do not result in a statistically significant rating difference, the compression is said to be lossless. Differences that do result in significant rating difference are said to be lossy.

The final element in the chain is the "information extractor." In this book, that is a trained observer. The observer may be a trained imagery analyst (IA) or a trained target acquisition professional such as a tank gunner or aircraft weapons officer. Although machine processing may be successful some time in the future, it has not been to date. The success of the extraction process depends on both perceptual and cognitive abilities of the observer.

1.3   Performance Prediction Approaches

A variety of approaches have been taken in modeling and predicting the performance of S&R systems. In general, these approaches are a mix of theoretical and empirical methods. Some require a physical image (image-based models), others require only a system design (parameter-based models). The latter, however, have generally been empirically developed and validated and can also be used with image measurements. Some approaches are designed to predict perceptual differences as opposed to absolute performance. In the sections that follow, each approach is briefly described. The models have been divided into four categories:

1. Physical image quality models;
2. Physical and image simulation models;
3. Information theory-based models;
4. Difference models.

Major emphasis is placed on the physical image quality models (both image- and parameter-based) as they are the most fully developed and validated. Physical and image simulation models are briefly discussed. Information theory–based models and difference models are also discussed in some detail.

1.3.1   Physical Image Quality

The most common form of S&R performance prediction is based on physical image quality measurements or predictions of those measurements. The ability to extract information from an image is based on the ability of the observer to detect objects of various types in a scene. This in turn is based on the ability to detect edges or tonal changes in the image. Large changes, large objects, and well-defined edges are easier to detect than those that are small and ill-defined. The presence of noise degrades detection performance. Consequently, measures that relate to the ability to see small detail and edges, as well as measures relating to contrast and noise, can be used to predict observer performance. Typical unitary measures include scale, resolution, contrast, and some noise-related measure. A measure of edge sharpness is also common. With the possible exception of image scale (ratio of image distance to object distance), each of these measures has been defined in a variety of ways. These measures are defined in detail in Chapter 6. Performance predictions using these measures may predict performance from a single measure; for example,

$$\text{Performance} = k + a \times \textit{Scale} \qquad (1.1)$$

where k is a constant, and a is a multiplicative coefficient or from multiple measures; for example,

$$\text{Performance} = k + a \times \textit{Scale} + b \times \textit{Resolution} + c \times \textit{Contrast} \qquad (1.2)$$

where k is a constant, and a, b, c are multiplicative coefficients.

More complex physical quality measures may also be used to predict performance. For example, the modulation transfer function (MTF) defines contrast as a function of spatial frequency and thus effectively combines both contrast and resolution. In (1.2), MTF could be measured using a physical target and an MTF-derived measure substituted for resolution and contrast. These physical quality-based approaches are models in the sense of regression models. They predict performance as a function of one or more independent variables.

Physical image quality models require a set of empirical data to develop an initial model. Once the model is validated, performance predictions can also be made on the basis of system design information. For example, system component MTFs (optics, atmosphere, etc.) can be cascaded to define a system MTF, and from that information and an existing performance prediction model, a prediction of system performance can be made. Similarly, estimates or knowledge of radar system impulse response and noise performance can be used to predict system performance. Although physical images are not always required for such predictions, it is generally necessary to validate the predictions after the system becomes operational. Again, these generally take the form of a regression model.

1.3.2 Physical Models and Simulations

A second class of models includes physical models and simulations. In the case of physical models, targets and target backgrounds are physically modeled to some relatively small scale [4]. For visible systems, sun and atmospheric effects can be optically and mechanically simulated and an image generated of the model. The image itself can be considered as a performance prediction, or it may be measured and the measurements used as input to another model. The images may also be viewed by human observers and their responses used to generate a performance prediction model.

In the case of IR and SAR systems, image simulation is more common than physical modeling. For IR imagery, the appearance of a target is a function of emitted as well as reflected energy—it is very difficult to model variations in emitted energy. For SAR imagery, the need for a coherent energy source coupled with the problem of scaling wavelengths creates a problem. Image simulations can be created by generating synthetic objects and backgrounds and using ray tracing to generate synthetic images. The synthetic objects and backgrounds are provided with the necessary reflective and emissive properties of the actual materials. This forms the synthetic scene model. At the opposite end of the simulation is the sensor model. Between the sensor and scene model is the energy transmission model. For an IR system, this would include radiance, meteorological, and atmospheric effects. For an SAR system, the transmitted and reflected radar energy would be modeled.

Physical and image simulation models have largely been used to understand imaging phenomena and as training aids. In these applications, fidelity is of a lesser concern. They have not been used in any large-scale fashion to predict overall system performance. For this reason, they will not be treated in subsequent chapters.

1.3.3 Information Theory–Based Models

The information theory-based approaches are based on the work of Shannon and Weaver in communication theory [5]. Shannon and Weaver defined information as a reduction in uncertainty (due to the content of a message). The basic unit of measure was the bit (binary digit). In the most simple form for an image, the information content of an image can be expressed as [6]

$$I = N \log_2 L \qquad (1.3)$$

where I = information content, N = number of display elements (pixels for a digital image), and L = number of response levels per display element.

Information density, or information content per unit area, can be used to compare different systems or processes. Information density is defined as

$$ID = I/A \qquad (1.4)$$

where *ID* = information density, *I* = information content, and *A* = image or display area.

This expression applies for discrete element displays, not to imagery as such. Note also that the number of response levels per display element is the number of levels that are discriminable by the observer. This is typically less than those that are theoretically present. For an S&R image, information density is defined as [7]

$$ID = \pi \int_0^\infty \log_2 \left(1 + \frac{P_S(r)}{P_N(r)} \right) r \; dr \qquad (1.5)$$

where $P_S(r)$ is signal power for frequency r, and $P_N(r)$ is noise power for spatial frequency r.

Information density is closely related to the power spectrum of the image [6]. Both optical and digital methods of deriving the power spectrum from an image have been used as a means of predicting performance. The power spectrum can also be derived from knowledge of system design. In either case, it is necessary to relate information density to some measure of perceptual performance (e.g., probability of identification).

1.3.4 Difference Models and Metrics

Difference models compare two images and relate one or more difference metrics to a difference in the ability to extract information or, in some cases, a difference in perceived quality. One of the most common metrics is the root-mean-square error (RMSE) metric. It computes the standard deviation of pixel amplitude differences between two images (often an original and a compressed or processed version of the same image). Peak signal-to-noise ratio (PSNR) is another common metric used to assess the effects of processing. Values for these metrics are then compared to perceptual metrics such as perceived degradation or subjective quality.

A more complex and perhaps more interesting model is the just-noticeable-difference (JND) model developed by Sarnoff [8]. This model or method computes a difference image based on a model of the human visual system. Differences between an input image and a simulated image based on a set of design parameters (e.g., bandwidth compression) are computed and displayed as an image where intensity is proportional to the probability that an observer would detect a difference (JND) between the two images. JND values can, in turn, be related to other perceptual measures, such as NIIRS difference (delta-NIIRS) or perceived degradation.

1.4 Summary

A variety of approaches have been taken in modeling and predicting the performance of S&R systems. They range from physical modeling to regression modeling to theoretical modeling. Most approaches have tended to concentrate on a limited number of elements in the overall image chain. No single approach has been shown to successfully handle all performance prediction applications. It is for this reason that multiple approaches are often necessary.

References

[1]   Leachtenauer, J., et al., "The General Image Quality Equation," *Applied Optics*, November 1997, pp. 8322–8328.

[2]   Ratches, J., *NVL Static Performance Model for Thermal Viewing Systems*, Report ECOM 7043, Fort Monmouth, NJ: USA Electronics Command, April 1975.

[3]   Leachtenauer, J. C., "National Imagery Interpretability Rating Scales: Overview and Product Description," *ASPRS/ASCM Annual Convention & Exhibition*, Vol. 1, April 1996, Baltimore, Maryland, pp. 262–271.

[4]   Schott, J. R., *Remote Sensing: The Image Chain Approach*, New York, NY: Oxford University Press, 1997.

[5]   Shannon, C. E., and W. Weaver, *The Mathematical Theory of Communication*, Urbana, IL: University of Illinois Press, 1949.

[6]   Schindler, R. A., and W. L. Martin, *Optical Power Spectrum Analysis of Display Imagery*, AMRL-TR-78-51, Wright-Patterson Air Force Base, OH: Aerospace Medical Research Laboratory, 1978.

[7]   Dainty, J. C., and R. Shaw, *Image Science-Principles, Analysis, and Evaluation of Photographic Type Imaging Processes*, New York, NY: Academic Press, 1974.

[8]   Lubin, J., *A Methodology for Imaging System Design and Evaluation*, Princeton, NJ: David Sarnoff Research Center, 1995.

2

Surveillance and Reconnaissance Imaging Systems

In this chapter, three types of imaging systems are described: visible (photographic and electro-optical), infrared (IR), and synthetic aperture radar (SAR). Image formation methods are also discussed, including line scanning, staring, and video (TV). Electro-optical (EO) and IR systems and the imaging processes are described, and the system parameters are presented. The characteristics of SAR are then described. Finally, some current S&R systems are provided as examples.

2.1 Introduction

There are many ways to group or categorize S&R imaging systems. By wavelength, the systems can be described as visible, IR , or radar. Systems operating in the visible portion of the spectrum can be described as photographic or EO. EO systems image in both the visible and IR portion of the spectrum. It is common, however, to restrict the term EO to those sensors operating in the visible through short-wave IR (SWIR) portions of the spectrum [1]. EO systems, as shown in Figure 2.1, can cover ultraviolet (UV), visible, near infrared (NIR), and SWIR wavelengths. UV systems are not typically used in S&R imaging, but visible systems are used frequently and cover the 0.4–0.7 μm region (the response of the human eye). Photographic systems cover the 0.4–0.7 μm region, but can extend to 1.0 μm. More commonly, an EO system [1] will respond to energy from both the visible region and extend past the NIR and into the SWIR. EO systems are grouped this way because

13

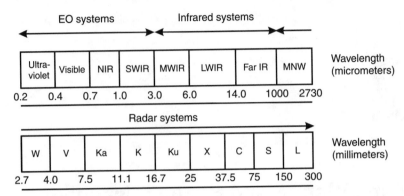

Figure 2.1 Systems and corresponding spectra.

most of the imaged energy is reflected from objects where the source of the energy is from the sun, moon, or stars.

Midwave infrared (MWIR) systems [2] are those that respond to energy from 3–5 µm, or some subband within. Midwave systems detect energy that can be object-reflected, but much of the energy is emitted from the objects according to Planck's radiation laws (except in the case where direct sunlight is reflected). Longwave infrared (LWIR) systems [3] respond primarily to energy emitted from objects.

Because most objects on Earth emit power at a peak wavelength of around 10 µm, LWIR has been the sensor of preference for U.S. Army target acquisition systems (TAS) for the past 40 years. The U.S. Navy has migrated to MWIR systems because the LWIR benefit is reduced by water content in the air in most maritime environments. MWIR staring focal plane arrays have matured in the past 10 years to where the MWIR performance is as good as LWIR performance in the emissive-only, or nighttime, environments. A scanned system suffers from short detector integration times compared to a staring sensor, so now the Army is currently evaluating MWIR systems for many applications. Also, most LWIR systems still use scanners and coolers that are associated with high power consumption and weight, but LWIR staring arrays are currently being fabricated in limited quantities. Once these arrays are matured and mass-produced, LWIR staring arrays will again be the sensor of choice for non-maritime terrestrial observations.

Radar wavelengths are longer than those found in the infrared. Also, radar systems provide their own illumination beam where the reflected beam is processed to form an image. Radar wavelengths are also described in Figure 2.1, where the most common are the Ka, X, L, and C bands.

In recent years, multi- and hyper-spectral imagers have been developed. Multispectral (MS) means that the sensor can acquire imagery in more than a single spectral band [4]. Three to five–band imagers have been constructed with multiple visible bands, NIR, and SWIR combinations. Also, different bands within LWIR and MWIR have been used to create an MS sensor. Effective MS sensors are also created by combining the output of two or more sensors (e.g., visible and MWIR or radar) using sensor fusion. Hyperspectral (HS) means that, at a minimum, hundreds of bands are acquired by a particular imager, and the imager covers a broadband with images created in tiny bandwidths. The spectral nature of the imagery is then exploited for some application, often for materials identification based on a spectral signature.

From Figure 2.2, it is easy to see which bands can be used in the Earth's atmosphere. Bands where the transmission of energy through the atmosphere is good are desirable. The light is not attenuated between the source object and the sensor. There are many good bands in the visible, NIR, and SWIR wavelengths. One would not, however, design a sensor in the 5–8 µm wavelength region because the light is absorbed and scattered by the atmosphere. Note that the midwave and longwave IR bands correspond to "windows" in the atmosphere. There are many windows in the electro-optical wavelengths.

Visible and IR sensors can also be grouped by the arrangement of their detectors and how the detectors are spatially dispersed over the image (i.e., image formation). Photographic systems use film as a detector. An image of a scene is focused on a film plane, and the response of the film to detected energy is used to form an image. For a variety of reasons, EO systems have replaced photographic systems for S&R applications. Line scanners and staring arrays are the most common EO and IR sensor configurations in

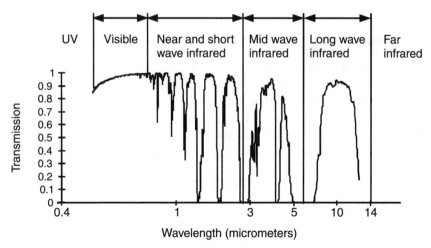

Figure 2.2 Atmospheric transmission of a 1-km horizontal path.

surveillance and reconnaissance systems. Line scanners are sensors with a line of detectors that are scanned across the image plane. Staring arrays require no scanner to collect an image, because a two-dimensional array of detectors is dispersed over the image plane. These systems are described in detail in the following sections.

2.2 Imaging with IR and EO Sensors

Many S&R platforms are equipped with both IR and EO systems. Predator, Global Hawk, and Landsat 7 systems are described as examples later in this chapter and all are equipped with both IR and EO sensors. Resolution is proportional to wavelength. Consequently, for a given set of optics, resolution is better in the visible wavelengths (EO) than in the IR. EO provides higher performance, however, only if sufficient scene illumination is provided by the sun or other sources. IR provides less resolution but performs in the absence of external illumination. Therefore, with platforms that are equipped with both, it is usually the case that IR is used at night and EO is used during the day. This concept is important because IR systems are usually analyzed in the context of night conditions where solar irradiance is not present. The performance of IR systems is usually presented in terms of the infrared emission of sources. A warning is appropriate here, because some IR systems are used in the daytime and solar reflections are seen prominently in the MWIR band. Given this warning, the imaging process for the IR and EO systems will begin with IR, because the single path of energy through the atmosphere makes the IR analysis simpler.

2.2.1 Infrared Sensors

The analysis of infrared sensor performance is not quite as complicated as the analysis of visible EO imaging systems. Consider the scene in Figure 2.3, where a helicopter sensor is imaging a tank in the IR band. The scene shown was taken in the LWIR band. The helicopter cartoon is shown to make a point. For most IR imaging systems, the primary use involves imaging at night or in the dark when little to no illumination is present. All objects above 0 Kelvin emit IR radiation in accordance with Planck's law [5]. Consider the emittance curves shown in Figure 2.4. These curves were generated using Planck's equation:

$$M(\lambda, T) = \varepsilon(\lambda) \frac{c_1}{\lambda^5} \frac{1}{e^{c_2/\lambda T} - 1} \quad [\text{W/cm}^2\text{-}\mu\text{m}] \qquad (2.1)$$

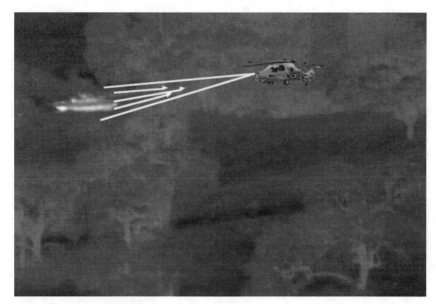

Figure 2.3 Imaging in the infrared.

where M is emittance, λ is the wavelength in μm, T is the source temperature in Kelvin, c_1 is a constant (3.7418×10^4 W-μm^4/cm^2), and c_2 is another constant (1.4388×10^4 μm-K). The $\varepsilon(\lambda)$ term is emissivity and describes the surface emissive characteristics of the source. A blackbody source is defined as a source with an emissivity of 1 over all wavelengths. Emissivity is a

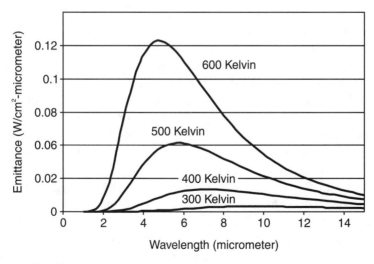

Figure 2.4 Planck's equation.

function of the degree to which a body absorbs and emits radiant energy. A highly reflective surface (e.g., polished aluminum) will absorb little energy and has low emissivity. From Figure 2.4, note that emittance depends on wavelength and source temperature. To calculate the power in a particular band, (2.1) is integrated over wavelength and the emittance is multiplied by source area. A few characteristics are worth mentioning here. First, the emittance varies dramatically with source temperature. In fact, Stephan-Boltzmann's law states that the integrated emittance over all wavelengths is

$$M(T) = \varepsilon \sigma T^4 \quad [\text{W/cm}^2] \tag{2.2}$$

where σ is the Stephan-Boltzmann constant 5.67×10^{-12} W/(cm²-K⁴). This equation only applies to blackbodies or graybodies where emissivity is constant (less than 1) over all wavelengths. The second characteristic is that the peak wavelength of emission shifts to the shorter wavelengths with high temperatures. Wein's law describes this peak wavelength:

$$\lambda_{pk} = \frac{2897.8}{T}. \quad [\mu\text{m}] \tag{2.3}$$

The peak wavelength of objects at terrestrial temperatures (300 K) is around 9.6 μm in the LWIR band. The sun is nearly a blackbody source with a 5600-K temperature with a peak wavelength of emission at 0.55 μm (in the center of the human visual band).

Because IR radiation is the self-radiation of thermal bodies, no illumination energy is required to image a scene; the energy is self-emitted by the objects in the scene. With no illumination path to consider, the analysis includes only the calculation of power from the target, the reduction of power by the atmospheric transmission and the collection and image reconstruction of the power by the sensor. This is a single atmospheric path process. The amount of power provided by the source depends on the temperature and emissivity (surface emission characteristics) of the object. This power is emitted in all directions, and only the rays from the source to the sensor aperture are collected for imaging.

In Figure 2.3, white corresponds to objects that are high in temperature and emissivity. The tank tracks are hot and highly emissive. Black corresponds to those objects that are cold, low in emissivity, or both. The sky and parts of the ground fall within these guidelines. Obviously, a tank that has been running and moving (i.e., hot tracks) is easier to detect because of the hot tank parts. Note that the process is really an optical power process that is primarily a function of temperature and emissivity.

Infrared sensors are used in S&R systems for a number of purposes. The primary purpose is S&R applications at night. These sensors, however, are used sometimes during the day to monitor the relative levels of temperatures in buildings, grounds, etc.

2.2.2 Electro-Optical Sensors

EO systems operating in visible wavelengths are more difficult to analyze than infrared systems, because the signal-to-noise analysis involves a signal that corresponds to a two-path propagation [6,7]. Consider the EO scenario shown in Figure 2.5. For most targets or objects of military interest, the self-emissive power is extremely small in the electro-optical wavelengths. The majority of available radiation comes from the reflective component of the target and background. The illumination source is typically the sun, moon, or stars. In the figure shown, the sun illuminates the tank. The spectral irradiance of the sun, however, is modified by the atmospheric path of the sun to the tank. This modification can vary dramatically depending on sun position and atmospheric conditions. The atmospheric conditions can range from a clear day with no clouds to heavy clouds and heavy rain. The modified irradiance is multiplied by the reflectance of the tank and/or background, and the light takes the path of the tank to the sensor. The reflectances, unlike emissivity in the IR bands, vary significantly over the electro-optical wavebands. It is these variations that provide color. These variations can cause

Figure 2.5 Electro-optical imaging.

an averaging effect that makes the background look very similar to the target. Once the light is reflected by the target/background, the light takes a second path to the sensor. This path must also modify the optical power and can change dramatically depending on path length geometry and atmospheric conditions. The analysis of this dual-path signal propagation is much more difficult and cumbersome than the single-path infrared system analysis.

2.3   Image Formation

The formation of an image can be accomplished a number of different ways. In this section, line scanners, staring arrays, and television systems are presented. Line scanners have been used extensively in S&R imaging. Staring sensors (also called focal plane arrays, FPA) are becoming more common, and analog television systems are becoming less common. Digital TV systems, however, may become more common in S&R applications.

2.3.1   Line Scanners

Line scanners are sometimes called "push-broom" scanner systems. A line of detectors is swept across the image in the same manner as a broom. Both EO and IR systems are realized as line scanner systems. Consider the line scanner system shown in Figure 2.6. A detector array is positioned at the focal length of an optical system (both the optics and the detector are part of the sensor). If the range is large, then an image of the ground is formed at the focal length of the optics. A line of detectors (linear detector array) is scanned across the image. The scanning can be accomplished by actually moving the array or by a variety of mirror types moving the image across the detector array. As the detector array traverses the image plane, a corresponding coverage is scanned on the ground. The angle described by the optical focal length and the detector array length gives the angle known as the field of view (FOV).

$$FOV = 2 \tan^{-1} \frac{d}{2f} \quad [\text{degrees}] \quad (2.4)$$

In Figure 2.6, the FOV for the cross-scan direction is shown. The FOV in the scan direction depends on the distance that the detector array is scanned on the image. The ground coverage in the cross-scan direction is shown and

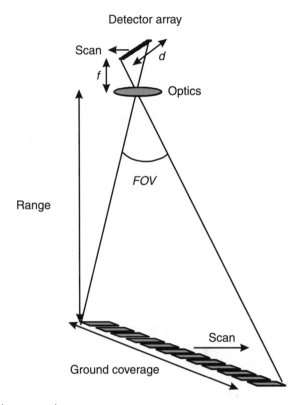

Figure 2.6 Line scanner imager.

can be determined by projecting the FOV to the range shown. The ground coverage in the scan direction, again, depends on the distance that the detector is scanned across the image plane. In addition, in some line scanner systems, the entire optics and detector assembly is rotated to give an extremely large scan direction angle. This technique is frequently called gimbal scan because a gimbal is required to move the entire assembly. Angles of 120° are easily achieved using gimbal scanning.

In some platforms, the scan direction is the direction normal to the direction of flight. In this case, the line scanner sweeps back and forth under the aircraft in wide-angle "swaths." Once the platform covers a ground distance defined by the along-track FOV, another swath is collected. In another configuration, the scan direction is the direction of flight and the flight motion provides the scanning action. Under these conditions, a large number of detectors are required in the linear array to provide good cross-scan coverage.

2.3.2 Staring Array Sensors

Staring array sensors are sensors with a two-dimensional detector array in the image plane (sometimes called the focal plane) of the imager. Because the detectors are distributed spatially over the image, a scanner is not needed in order to obtain a "frame" of imagery (see Figure 2.7). The advantage of a staring array is increased sensitivity. A detector in the array is allowed to integrate a point in the image for a much longer duration (even up to the frame rate of the sensor). For example, if an image is obtained every 1/60th of a second, a detector can integrate a point on the image for up to 1/60th of a second. For a line scanner imager to obtain the same image in 1/60th of a second, the detector must scan across the image so that a detector can integrate only a point in the image for a small fraction of the frame time.

There are two primary disadvantages for staring array (FPA) imagers. The first is a problem with image sampling. In a staring array, a detector can only be as large as the spacing between detectors. In typical scanned imag-

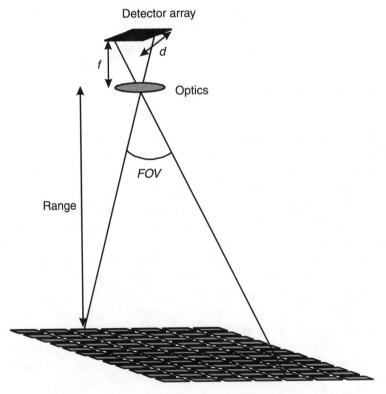

Figure 2.7 Staring array imager.

ers, image samples can be taken at intervals smaller than the detector. For systems with small optics blur and small detectors, the coarser image sampling leads to aliasing, or image artifacts due to signal corruption. Aliasing and sampling artifacts are described in Chapter 4. The second disadvantage is that for large area coverage, the staring imager must be "stair-stepped" across the area of coverage. It does not allow for the smooth, flowing imagery collection associated with a line scanner. Stair-stepping design can be complicated, and the stitching of imagery from different snapshots can be a difficult image processing task.

Overall, the disadvantages of a staring array are outweighed by the large increase in sensitivity associated with long detector integration times. The aliasing can be controlled with good optical design or microscanning (where the focal plane is moved in small steps between frames) techniques.

2.3.3 Television Sensors

Television systems [8] are EO systems (although some research was performed on IR television systems) with camera tubes that act as light detectors. The television tube converts the optical signal to an electrical signal. The vidicon, shown in Figure 2.8, is the most widely used camera tube. The vidicon uses an electronic beam to scan a photoconductive target that acts as the light sensor. The target is a transparent conductive layer that is applied to a photoconductor. The signal electrode is operated at a positive voltage with respect to the back of the photoconductor that operates at the cathode voltage. The scanning beam initially charges the back of the target to the cathode voltage. When a light image is formed on the target, its conductivity

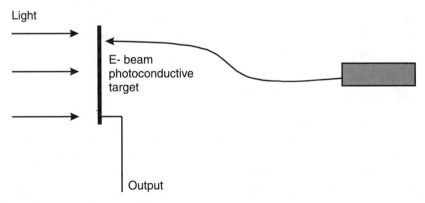

Figure 2.8 Television camera tube.

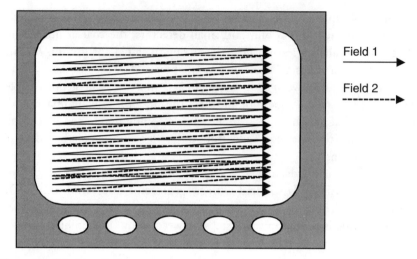

Figure 2.9 Television display.

increases in the illuminated areas and the back of the target increases in potential. The signal is read out as the electron beam traverses the target.

The electron beam traverses a pattern that matches the pattern written on a television display. A field is a signal that has been serially written on a television display at a rate of 59.9 Hz (for U.S. television in the NTSC format [9]). Two fields are interlaced to give a frame (29.95 Hz) that has 525 television lines. Because of the time required, however, to move the beam from line to line and field to field (flyback and retrace, respectively), the effective number of lines is ~490. Each line is written across the display and retraced to the left of the display (Figure 2.9).

Television systems made up the majority of EO systems until the solid-state detector staring array was developed. In particular, charge coupled devices (CCDs) in the form of staring arrays currently dominate the S&R electro-optical system area.

2.3.4 Analog versus Digital Signals

The television system described above is an example of an analog process. The light that is detected is continuous, the generated voltage is continuous, and the displayed luminance is continuous. In many S&R systems, the signal would have been recorded or transmitted over a data link in a continuous manner. Unfortunately, analog signals are easily corrupted with limited bandwidth electronics and noise sources in the signal chain.

Many sensors today are digitized prior to the recording or transmission of the signal. The output of a staring array (e.g., a charged coupled array) can be analog if the charge is read out in a serial manner. The output can also be digital if the charge is quantized by a digitizer on either the detector chip or the readout circuit. IR line scanners are typically digitized on the readout electronics card (off the detector substrate). Future IR sensors are anticipated to have on-chip digitizers (readout integrated circuit, ROIC). The benefit of the digital signal is that the information can be transmitted with little (if image compression is used) or no degradation. The disadvantage is that the quantization process can be considered a noisy process that, if not correctly addressed, can limit the performance of the system.

2.4 Imaging System Parameters

There are a large number of imaging system parameters that make up a system and describe the system performance. These parameters range from the general parameters of sensitivity and resolution to parameters that are more specific. While there are hundreds of specific imaging system parameters, we describe a few of the more important ones in this section. We begin, however, with the two general parameters of any imaging system: sensitivity and resolution. Many of these parameters are discussed again in Chapter 6 in the context of performance predictor variables.

2.4.1 Sensitivity and Resolution

All imaging systems have the general performance characteristics of sensitivity and resolution. Sensitivity [10] describes how noise in an image will appear at the output of the sensor. Resolution describes how fine spatial dimension in the scene will appear.

As stated previously, the signal-to-noise ratio in an IR system depends on the signal from the source, where the signal is a function of variations in temperature and emissivity. The signal is reduced as it propagates through the atmosphere, where absorption and scattering occur. The signal is collected by the sensor and imaged onto a detector array. The noise associated with the detection process is usually background limited photo detection (blip) noise. The ambient temperature and emissivity in the scene provide discrete photon events to occur at the detector, and a large number of these photon-absorption, electron-transition events cause the signal to be accompanied by noise. In a lower temperature environment, the signal is not accompanied by a large number of background photons (which cause noise), so higher sensitivity can be achieved. The sensitivity of an IR imager is sometimes described as "blip-limited." Because shorter wavelength photons have more energy, there are a smaller number of photons that make up an equivalent signal, thus reducing the overall noise in the signal detection process.

For EO systems, the primary noise source at low light levels is electronic readout noise that is a collection of reset noise, white noise, and $1/f$ noise. The reset noise is the Johnson's noise associated with the resetting of the sensing capacitor. The white noise is a collection of Johnson noise sources in the electronics along with the amplifier noises associated with on- and off-chip amplifiers. The $1/f$ noise is caused by a number of fabrication artifacts, including contact manufacturing. At high light levels, the EO system sensitivity can be limited by photon, or shot, noise corresponding to the discrete nature of photon detection and fixed-pattern noise that is associated with the different responsivities of the detectors.

The resolution of a particular sensor is usually limited by one of three characteristics: the quality of the optics, the detector size, or the sample spacing. The longer wavelengths in the infrared give rise to larger diffraction spots. IR systems are usually diffraction-limited, meaning that the size of the diffraction spot is larger than any other component impulse response in the system. Because visible wavelengths are so short, the diffraction spot is small, so the optics of these systems are usually aberration-limited. The finite size of the detector can limit the resolution of both IR and EO systems. Larger detectors give low resolution and small detectors give high resolution. Finally, if the detectors are not spaced closely together, or sampled in small spatial increments as they traverse an image, then the sample spacing is gross and can limit the resolution of the system.

IR detectors tend to be larger in dimension than EO detectors. Overall, an EO sensor, with equivalent optical dimensions and package size can have a significantly higher resolution than that of an IR system. Under low light conditions, however, the sensitivity of the IR system can be higher than that of the EO sensor. Later in this chapter, we describe some general guidelines for choosing sensor band.

2.4.2 Field of View

The FOV as given by (2.1) is of primary importance in any S&R system (see Figure 2.10) [11]. A wide FOV can give larger area coverage, but generally has lower resolution. The detectors are spread over a larger coverage area. In a narrow FOV, the detectors are spread over a small area, thus increasing the resolution and sacrificing area coverage. It is very difficult to detect an object with a narrow FOV. It is sometimes likened to "looking through a soda

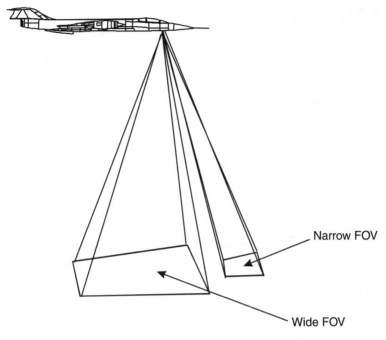

Figure 2.10 Narrow and wide FOVs.

straw." The selection of an FOV is a tradeoff between area coverage and resolution. The best design gives just enough resolution to achieve the desired task and, therefore, maximizes area coverage.

2.4.3 Field of Regard

Field of regard (FOR) is similar to FOV, but describes the angle of coverage that a sensor is capable of, given an infinite amount of time. A gimbaled sensor may have an FOR of 120 degrees by 90 degrees, but only have an FOV of 2 degrees by 2 degrees. For this example, it may take a great deal of time for the sensor to cover the entire FOR with the given FOV. Note that the FOR is important in that it describes the capability of a sensor to view certain areas given the platform geometries and conditions. In Figure 2.11, the FOV of the sensor is limited only to the areas forward of the aircraft, since the FOR does not extend to positions behind the aircraft.

2.4.4 Diffraction

Diffraction has been frequently described as providing the fundamental resolution limit of an EO/IR imaging sensor. Diffraction is the generation of

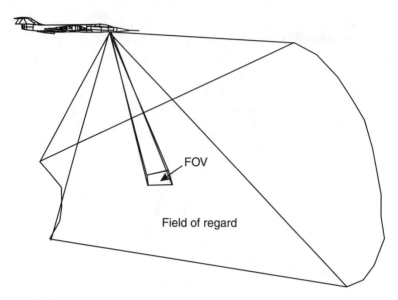

Figure 2.11   Field of regard.

secondary wavefronts as energy passes by an edge or through an opening. These secondary wavefronts interfere with the primary wavefronts as well as with each other to form a diffracted wave. In simple terms, when energy passes through a small opening, the light bends around the edges. The smaller the opening or the longer the wavelength of light, the more bending occurs. For EO and IR imaging systems, this bending manifests itself as a blur on the image plane. This blur can be projected in angular space in front of the sensor and can even be projected onto the ground in front of the sensor. Consider Figure 2.12. Energy propagates from the ground to the imaging optics. The obstructing aperture is the imaging optics diameter, where the imaging optics form an image one focal length away at the image plane. A small point on the ground causes a diffraction blur at the image plane. This blur can be projected in angle in front of the sensor so that the diameter (to the first zeros) of the spot are $2.44\lambda/D$, where D is the diameter of the imaging optics and λ is the wavelength of the energy. Note that the diffraction blur spot can even be projected onto the ground by multiplying by range, R. This blur can be compared to the object that is being imaged to see if sufficient resolution is provided by the sensor optics (due only to diffraction) to discern necessary details. Recall that a point source on the ground causes this blur, not a blur spot on the ground. The projection of the blur spot to the ground is just a method for comparing the diffraction blur effects to the

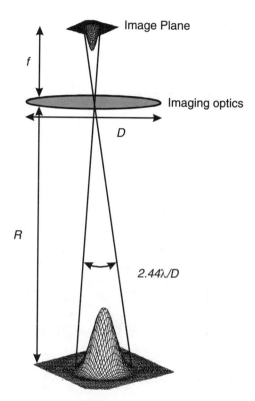

Figure 2.12 Diffraction in imaging systems.

objects that will be imaged. If the objects are significantly less in size than the projected diffraction blur, the sensor will not be able to adequately image the object. If the diffraction blur is significantly smaller than the object size, there is a good chance that the sensor will be able to resolve good detail on the object.

2.4.5 Detector Angular Subtense and Instantaneous Field of View

Whether a sensor is scanned or is a staring array, the detector angular subtense (DAS) or instantaneous field of view (IFOV) is a description of the sensor's detector shape (a single detector in the detector array) as projected, in angle, in front of the sensor. If a point on the ground is within this DAS, then the same signal is seen on the output of the detector. Therefore, this projected angular subtense can be a quantity that describes the resolution associated with the detector size. From Figure 2.13, the DAS is

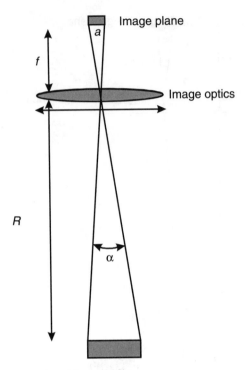

Figure 2.13 Detector angular subtense (DAS).

$$\alpha = \frac{a}{f} \quad \text{[milliradians]} \tag{2.5}$$

where a is the detector dimension and f is the focal length.

A small detector (and small DAS) can resolve high spatial resolution, and a large detector (and large DAS) gives poor spatial resolution. Also, two DASs are usually provided for a given sensor, one in the horizontal (scan) direction and one in the vertical (cross-scan) direction. Finally, IFOV has come to mean the same thing as DAS and is usually given in milliradians. Long ago, IFOV was the solid angle, in steradians, of the detector projected forward of the imaging lens. While this was a good measure for signal collection, it does not describe the horizontal and vertical resolution associated with the detector size.

2.4.6 Ground Sample Distance

The ground sample distance (GSD) is a measure of resolution limitations due to sampling [12]. In Figure 2.11, a detector array is exaggerated with only

12 detectors to show the concept. The detector pitch, p, is the distance between detector centers. For a scanned sensor, the pixel pitch is the distance between points where the output of the detector is sampled as the detector is scanned across the image plane. The pixel pitch divided by the sensor focal length gives the angular distance between sensor samples. If this angular distance is projected to the ground, it defines the GSD (Figure 2.14):

$$GSD = \frac{p}{f} R \quad \text{[meters or feet]} \quad (2.6)$$

where p is detector pitch, f is focal length, and R is range.

The GSD can be compared to the projected DAS or the diffraction spot size to determine whether the system is limited in resolution by the diffraction effects, detector size, or the sampling (i.e., detector pitch). GSD can be a little more complicated in that if the FOV of the sensor is not directed

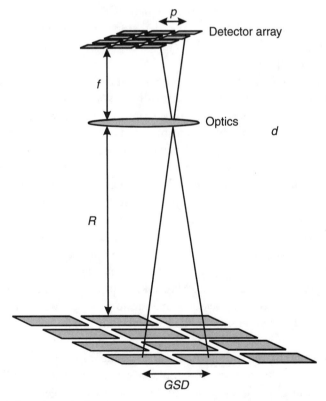

Figure 2.14 Ground sample distance.

normal to the ground surface, the GSD must be corrected for the look angle (the angle between the ground and the sensor line-of-sight):

$$GSD = \frac{pR}{f \cos \theta} \quad \text{[meters or feet]} \quad (2.7)$$

where p, f, and R are as in (2.6), and θ is the look angle.

Note that there is a horizontal and vertical GSD, but there is a two-dimensional system GSD that is taken as the geometric mean of the horizontal and vertical ground sample distances.

2.4.7 Point Spread Function

The point spread function (PSF) [13] is the output of an imaging system that views a tiny point (or impulse) of light. An infinitesimal size point on the input of the system creates a blur function on the output of the system. The PSF is a measure of the system blur, which includes the blur contributions from diffraction, optical aberrations, detector shape and motion, electronics, and display. In fact, the overall system blur is the convolution of each of the system's component blurs.

Usually the PSF takes on the general shape of the blur that corresponds to the limiting resolution of the sensor. If the system is diffraction-limited, then the PSF looks more like the diffraction pattern shown in Figure 2.12. If the system has a small diffraction blur and the system is display-limited, then the PSF takes on the general shape of the display blur.

The modulation transfer function (MTF) is the Fourier transform of the PSF. Resolution can be addressed in the spatial domain (milliradians) or in the spatial frequency domain (cycles per milliradian). They are inversely related, in that large spatial objects have significant low frequency content and small spatial objects have significant high spatial frequency content. We address the PSF/MTF relationships in detail in Chapter 4.

For digital S&R systems that store imagery that is later interpreted on various displays, sometimes the PSF is taken only as those blur components prior to image digitizing. This blur includes diffraction, aberrations, detector shape, and system vibration. This overall PSF size and shape in angular space can be compared to the sample spacing (the pitch divided by the focal length) in order to determine whether the system is adequately sampled. For systems that have three or more samples across the size of the PSF, the system is well sampled. For systems with less than two samples across the PSF, the system is undersampled and the image is corrupted by *aliasing*. Aliasing

Figure 2.15 (Left) Original scene, (middle) well sampled (blur is larger than sample spacing), and (right) undersampled (with aliasing).

is a sampling artifact that causes lines to be broken, shifted in space, and wider or narrower than in the scene. Figure 2.15 shows an input scene, a blurred or filtered scene, and an undersampled scene. The two scenes on the right have the same number of picture elements.

2.5 Synthetic Aperture Radar

SAR has become a common sensor in environmental monitoring, Earth resource mapping, and military systems [14–17]. SAR sensors provide the benefit of S&R global applications without severe degradation due to weather conditions. Both EO and IR systems can be rendered useless under heavy cloud cover. SAR, however, has the advantage (and disadvantage) of being "active," and providing its own illumination. The wavelength of the illumination (centimeter waves) travels through the atmosphere with little degradation. Although the wavelength is much longer than EO and IR system wavelengths, the (synthetic) aperture is also much larger, resulting in comparable resolution in azimuth. In range, resolution depends on bandwidth and can also be comparable to EO and IR.

In Figure 2.16, the aircraft points the beam to the side of the aircraft (side-looking radar). SAR is usually pointed perpendicular to the vehicle velocity vector. Typically, SAR produces a two-dimensional image. The first dimension, x, is called range and is a measure of the line-of-sight distance from the radar to the target. Range is determined by precisely measuring the time from the transmission of a pulse to the receipt of a target reflection. Range resolution in its simplest form is determined by the pulsewidth. The other dimension, y, is called azimuth, or cross-track, and is perpendicular to range. SAR has an inherent ability to differentiate fine azimuth resolution, which makes it very different than other radars. Instead of a huge antenna that focuses a small beam, the information is processed as the vehicle passes the target. Because the airborne radar collects data while flying, the data is

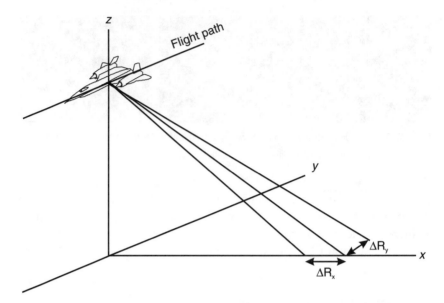

Figure 2.16 Synthetic aperture radar resolution.

processed as though the information came from a physically long antenna. Determining azimuth resolution can be described from a Doppler shift approach. A target position along the flightpath determines the amount of Doppler shift. Azimuth resolution corresponds to how finely Doppler shifts can be determined.

The range resolution, ΔR_x, is determined by controlling the illumination pulsewidth or encoding it to determine the location of the reflective surface. The y-dimension or azimuth resolution is determined by the beamwidth. In a real radar system, the beamwidth is inversely proportional to the antenna length. In a SAR system, however, the beamwidth is synthetically enhanced by using the incremental Doppler shift between adjacent points on the ground. Processing the signals converts a wide beam to a narrow beam, thus giving a high resolution effect. Also, the Doppler processing yields the same resolution on the ground regardless of the sensor-to-target range.

The resolution in the x-direction is determined by the pulsewidth time, τ, of the radar:

$$\Delta R_x = \frac{c\tau}{2} \quad [\text{meters}] \qquad (2.8)$$

where c is the speed of light (3×10^8 m/sec). The pulsewidth time is reduced to less than the actual pulsewidth with pulse compression techniques. There are a number of pulse compression methods that are used today.

Resolution in the y-direction, as stated above, is determined by the Doppler processing of the data. If time-independent filters are used in the processing, the *unfocused* case is realized. The Doppler shift in this case is determined by a signal over a time, T:

$$\Delta f = \frac{2V^2 T}{\lambda R} \quad [\text{Hz}] \tag{2.9}$$

where R is the range from the platform to the target and V is the aircraft velocity. If the Doppler shift is matched to the duration of data,

$$VT = \sqrt{\lambda R/2} = 0.707\sqrt{\lambda R} \quad [\text{m}] \tag{2.10}$$

where VT represents the distance flown during one filter time constant. This VT is equal to the resolution in the y-direction. The limit to resolution is in the processing. The longest length of time that data can be acquired is related to the beamwidth on the ground:

$$bw = \frac{R\lambda}{l} \quad [\text{m}] \tag{2.11}$$

where l is the antenna length. The time that data can be collected from a point on the ground is related to this beamwidth and the aircraft velocity:

$$t = \frac{R\lambda}{Vl}. \quad [\text{sec}] \tag{2.12}$$

This duration of time can be used in (2.8) to determine the Doppler frequency excursion:

$$\Delta f = \frac{2V}{l} \quad [\text{Hz}] \tag{2.13}$$

and the distance traversed during one filter time constant becomes

$$VT = \Delta R_y = l/2 \quad [\text{m}] \tag{2.14}$$

so the resolution limit associated with the azimuth direction is related to the antenna length. For "spotlight" cases where the beam direction is controlled on target for a longer period than a side-looking radar, the resolution can be higher.

The signal-to-noise ratio (SNR) of a SAR system is given by

$$S/N = \frac{PtG^2\lambda^2\sigma}{(4\pi)^3 R^4 (kT)L} \quad [\text{unitless}] \tag{2.15}$$

where t is the time on target, G is the antenna gain, σ is the radar target cross section, k is Boltzmann's constant (1.3807×10^{-23} J/K), T is the equivalent receiver noise temperature, and L is the transmit-receive loss. P is the average power

$$P = P_{pk}\tau PRF \quad [\text{Watts}] \tag{2.16}$$

where P_{pk} is the peak power, τ is the pulsewidth, and PRF is the pulse repetition frequency. The radar cross-section can be calculated as

$$\sigma = \Delta R_x \Delta R_y \sigma_o \quad [\text{m}^2] \tag{2.17}$$

where σ_o is a measure of surface roughness that ranges from 0.001 to 1 depending on angle of incidence, wavelength, polarization, etc.

A disadvantage of SAR is that returns are specular. Objects in a scene behave as mirrors and either reflect energy back to the radar receiver or reflect it away from the receiver. Surfaces with a high radar cross-section give very high signals, whereas those with a low radar cross-section give low signals. SAR imagery is less literal than EO imagery and thus may be more difficult to interpret initially. Imagery analysts go through a training to become proficient. An SAR image of a military vehicle in Figure 2.17 shows the nonliteral nature of SAR imagery.

Each pixel in an SAR image represents the intensity of the radar energy reflected back to the receiver. Energy may be reflected directly or it may bounce off multiple surfaces before reflection. Reflections from surfaces smaller than the radar resolution element may interfere both constructively and destructively to produce what is called clutter. Clutter appears to be noise-

Figure 2.17 SAR image of a military target.

like but varies in intensity as a function of surface roughness and dielectric properties. Clutter intensity or backscatter for pavement is lower than for grass; and that for grass is lower than that for trees. As a rule-of-thumb, high areas of backscatter correspond to rough surfaces. Flat surfaces, such as roads and freeways, usually reflect little energy and appear dark. Vegetation is moderately rough and appears gray. Surfaces inclined towards the radar appear brighter because the backscatter is higher. Also, when city streets and buildings are lined up in a manner that causes a "corner cube" double or triple bounce, the pixel appears very bright.

2.6 Choice of Imaging Sensors

It is difficult to generalize the selection of an imaging sensor for S&R purposes. An example is given in Figure 2.18 for the air-to-ground S&R application. If heavy cloud cover or rain is present, neither EO nor IR sensors are effective, so SAR is the sensor of choice (note that SAR will not penetrate thunderstorms). IR can be effective through light clouds or some fog, but not through heavy cloud cover with few breaks in the clouds. Also, if the application is long range, recall that SAR has a constant ground resolution with range. The ground resolution (ground sample distance and impulse response projected to range) of an EO or IR sensor degrades with range. SAR does not generally have as high a resolution as EO and IR sensors at short ranges, but can give sufficient resolution for the identification of military vehicles. Therefore, SAR may be a better choice of sensor for long-range standoff applications.

Given no heavy cloud cover and a range that is compatible with EO and IR area coverage and resolution, then EO and IR imagers are desirable

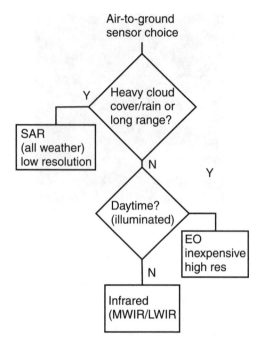

Figure 2.18 Air-to-ground imaging sensor selection.

because of their higher resolution and because EO and IR images are easy to interpret by imagery analysts. These images look very similar to images that we see with our eye every day. If sufficient external lighting is available, such as sunlight, then the EO system is the system of choice. The shorter wavelengths of EO systems mean that the diffraction blur is smaller than in IR systems. EO systems have been developed with commercial market help, so the detector technology gives smaller detectors with high efficiency at a tiny fraction of the cost of IR detectors. Therefore, EO systems are usually inexpensive compared to IR systems. In addition, they provide high resolution and sensitivity.

An IR system is required for use at night (or both day and night) where there is little to no external source of energy. An exception is for extremely short ranges (less than a few hundred meters), where an image intensifier may provide satisfactory performance. IR sensors have been shown to be effective for military vehicle S&R applications at ranges up to 20 km.

The most difficult choice is one of MWIR sensor versus LWIR sensor. This choice is usually accompanied by an analysis that includes the source characteristics, the atmospheric transmission, and the detector sensitivity. Terrestrial objects usually have more power in the LWIR region and hotter objects shift towards the MWIR region. A cool, dry atmosphere provides a benefit for the longwave band of up to a factor of 3, but a hot, wet atmosphere provides no significant benefit for longwave over midwave. Detector characteristics currently favor MWIR sensors. The sensitivity is much higher for midwave sensors. Focal plane arrays are readily available in the midwave, and provide large signal integration times.[1] Given the source, atmosphere, and detector characteristics, it is no surprise that the U.S. Navy relies heavily on midwave IR sensors, because they operate frequently in a hot, wet atmospheric environment.

The decision process that we have provided is oversimplified, but it gives the flavor of the decision process. There are many more accompanying issues that must be resolved such as cost, space, weight, and power requirements. Any one of these issues can alter the decision dramatically.

2.7 Examples of Surveillance and Reconnaissance Sensors

A number of S&R sensor and platform examples are provided in this section. These examples include unmanned aerial vehicles (UAVs) such as Predator and Global Hawk and satellite sensors such as Landsat, Radarsat, Ikonos, and KVR1000. The number of commercially available satellite systems is expected to increase substantially in the next few years; a listing of other systems is provided.

2.7.1 Predator

The Predator system is a UAV that was developed, beginning in 1994, as a Department of Defense Advanced Concept Technology Demonstration (ACTD) [18]. Since that time, Predator has flown in five combat area deployments. The aircraft is shown in Figure 2.19. The system includes an EO/IR payload, a SAR payload, satellite control, and both a Global Positioning System (GPS) and an Inertial Navigation System (INS). Predator can stay on station for 24 hours at a range of 400 nautical miles. The aircraft has a wingspan of 48.7 ft and a length of 27 ft with a 450-lb payload capacity. The Predator is currently in production by Aeronautical Systems Inc. and is deployed with the U.S. Air Force 11th and 15th Reconnaissance Squadrons.

Predator's EO/IR payload is manufactured by Versatron and includes a TV camera and an MWIR imager. The TV is a CCD camera and includes

---

1. Focal plane development is rapidly progressing in the longwave band.

Figure 2.19   Predator surveillance and reconnaissance system (Courtesy of Aeronautical Systems Inc.).

two modes: a zoom lens and a spotter scope. The zoom lens has a 10:1 zoom; focal length varies from 16–160 mm and the FOV varies from 2.3 by 1.7 degrees to 23 by 17 degrees. The spotter mode gives a 955-mm focal length and a FOV of 0.38 by 0.29 degrees. The IR camera is one of two staring array options, a 512 by 512 platinum silicide detector or a 256 by 256 indium antimonide detector. Both operate in the midwave infrared and include a Stirling cycle cooler. The PtSi option comes with a selectable six-FOV option that ranges from 1.4 by 1.0 degrees to 41 by 31 degrees. The InSb option also comes with selectable six FOVs ranging from 0.8 by 0.8 degrees to 41 by 41 degrees. The EO/IR payload includes a laser rangefinder from one of two laser options. One is an eye-safe Erbium glass laser that operates at 1.54 μm and the other is a Nd:YAG laser that operates at 1.06 μm. The Erbium laser has a range of 15,000m and the Nd:YAG laser has a range of 9995m. Both lasers have a resolution of 5m.

The Tactical Endurance Synthetic Aperture Radar (TESAR) is Predator's SAR system that is manufactured by Northrup Grumman. TESAR is a strip-mapping SAR that provides continuous variable-resolution from 1.0–0.3 m (1 ft). The strip map gives a swath width of 800m at a slant range of 4.4–10.8 km. The imagery is processed and displayed in a scrolling manner on an SAR workstation. TESAR has another mode called spotlight, where a selected region is continuously mapped as the aircraft moves past the region. With resolution of 0.3m, the spotlight provides an 800 by 800 m region.

With a resolution of 1 by 1 m, the spotlight provides a region of 2400 by 2400 m. A second SAR has been provided as an option on Predator, the Lynx SAR. The Lynx SAR has a 0.3–3.0m resolution stripmap mode and a 0.1–3.0m spotlight mode.

2.7.2 Global Hawk

Global Hawk [19] is another ACTD, but it was developed by the Defense Advanced Research Projects Agency (DARPA) and the Defense Airborne Reconnaissance Office (DARO). Global Hawk (Figure 2.20) is a high-performance UAV with a loiter altitude of 65,000 ft and a range of 13,500 nautical miles. The UAV is controlled by satellite link and can retrieve data over a number of direct or satellite links. The aircraft (Figure 2.20) is 44-ft long and has a wingspan of 116 ft. There are three sensors onboard: EO, IR, and SAR. The EO/IR sensor payload is packaged together as shown in Figure 2.21. The EO sensor has a response over a bandwidth of 0.4–0.8 µm. The IR sensor is a MWIR sensor with a 3.6–5.0 µm bandwidth. The SAR sensor can be used in a wide area search mode or a spot collection mode. The wide area search mode gives a 10-km swath that is collected as the aircraft flies over an area. The EO sensor is specified to provide a visible NIIRS of 6.5 at a 45-degree elevation angle (28-km range) and the IR sensor is specified to provide an infrared NIIRS of 5.5 at a 45-degree elevation angle (28-km range). The wide area search can cover 40,000 square nautical miles in one day, and the spot mode can collect 1900 2 by 2 km frames a day.

Figure 2.20 Global Hawk aircraft (courtesy of Raytheon).

Figure 2.21   Global Hawk EO/IR sensor package (courtesy of Raytheon).

The Global Hawk SAR is an X-band system (600 MHz) that can be operated in a 2.7.2 Global Hawk) mode with a 1-m resolution or a spotlight mode with a 0.3-m resolution. The WAS swath is a 10-km-wide strip, and the spotlight is a 2 by 2 km area. The sensor can also be used as a moving target indicator (MTI) in a manner very similar to the U.S. Air Force's JSTARS sensor. In this mode, moving vehicles cause a Doppler shift that is detected, and only objects that are moving are processed. The MTI mode can search a region from 20–200 km away from the platform or an area of 15,000 km^2 in one minute. The minimum detectable velocity is 4 knots. The SAR antenna is shown in Figure 2.22.

2.7.3 Landsat 7

Landsat 7 [20] is a remote sensing satellite that was launched on April 15, 1999, as part of the Landsat program. The mission of the Landsat program is to extend and improve upon the more than 26-year record of the Earth's continental surfaces provided by earlier Landsat satellites. Landsat 7 provides essential land surface data to a broad and diverse community of civil and commercial users. Landsat imagery was used to produce the first composite multispectral mosaic of the 48 contiguous United States. Landsat has also provided important information for monitoring agricultural productivity, water resources, urban growth, deforestation, and natural changes due to fire

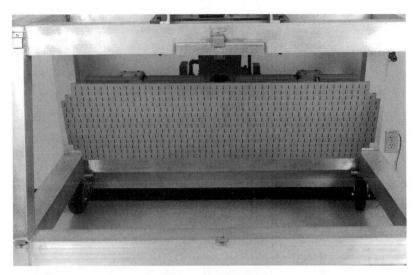

Figure 2.22 Global Hawk SAR antenna (courtesy of Raytheon).

and insects. Landsat imagery has also been used for mineral exploitation, to measure forest cover at the state level, and to monitor strip mining.

The sensor onboard the Landsat 7 satellite is called the Enhanced Thematic Mapper Plus (ETM+). The spacecraft was built by Lockheed Martin Missiles and Space Company in Valley Forge, Pennsylvania, and the sensor was built by Raytheon in Santa Barbara, California. The sensor is a passive EO/IR sensor with seven bands, as indicated in Table 2.1. The resolution described corresponds to the DAS of the detectors. A photograph of the sensor is shown in Figure 2.23.

Table 2.1
ETM+ Sensor Bands

Band	Resolution (m)	Spectral range (μm)
1	30	0.450–0.515
2	30	0.525–0.605
3	30	0.630–0.690
4	30	0.750–0.900
5	30	1.550–1.75
6	60	10.4–12.5
7	30	2.09–2.35
Pan	15	0.520–0.900

Figure 2.23   ETM+ (courtesy of NASA Goddard Space Flight Center).

The ETM+ is a push-broom scanning radiometer that covers a 183-km swath while orbiting at an altitude of 705 km (see Figure 2.24). The temporal resolution (revisit time) of the satellite is 8 days. The size of an imagery product is 183 by 170 km, and the system is capable of collecting 100 im-

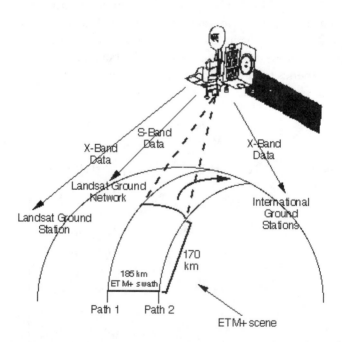

Figure 2.24   Landsat 7 coverage (courtesy of NASA Goddard Space Flight Center).

Figure 2.25 Landsat 7 image of San Francisco (courtesy of NASA Goddard Space Flight Center).

agery products per day. The imagery is available to customers within 24 hours of data acquisition. The Landsat 7 satellite is primarily intended for a 5-year design life. An example image of the Landsat 7 satellite is shown in Figure 2.25. The image is only part of an imagery product and shows the area of San Francisco.

2.7.4 RADARSAT

RADARSAT [21] is a Canadian-led project involving Canada and the United States. RADARSAT-1 is a satellite that carries an SAR sensor for obtaining detailed images of the Earth. It is used to monitor the Earth's resources and environmental changes and to support commercial industries such as fishing, shipping, agriculture, and oil and gas exploration. RADARSAT-1 has a planned lifetime of 5 years. The sensor uses a single-frequency, C-band transmitter that can steer the beam over a 500-km range. There are a wide variety of beam selections that can image swaths from 45 to 500 km in width with resolutions from 8 to 100 m. Incidence angles can range from 10 to 60 degrees. RADARSAT-1 repeats the same orbital path every 24 days; however, complete coverage at equatorial latitudes can be accomplished every 6 days using the 500-km swath width. The satellite is in a sun-synchronous polar orbit that places the satellites solar arrays in almost continuous sunlight. The satellite orbits at an altitude of 798 km with an inclination of 98.6 degrees. The satellite orbits the Earth 14 times per day and each orbit takes 100.7 minutes. The different imaging modes are shown in Table 2.2. The number of beam positions is within the incident angle range given. An example RADARSAT image is shown in Figure 2.26.

Table 2.2
RADARSAT Image Modes

Mode	Resolution (m)	Number of beam positions	Swath width (km)	Incidence angle (degrees)
Fine	8	15	45	37–47
Standard	30	7	100	20–49
Wide	30	3	150	20–45
Scan SAR Narrow	50	2	300	20–49
Scan SAR Wide	100	2	500	20–49
Extended (H)	18–27	3	75	52–58
Extended (L)	10–22	1	170	10–22

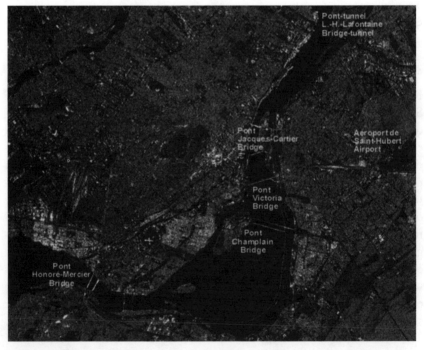

Figure 2.26 RADARSAT image of Montreal (courtesy of the Canadian Space Agency).

2.7.5 IKONOS

IKONOS [22] is a commercial satellite that was launched by Space Imaging, Inc. from Vandenburg Air Force Base on September 24, 1999. Lockheed Martin built the Athena II rocket that propelled the satellite into space. The camera system for IKONOS was designed and built by Kodak and can image a panchromatic scene at 1-m resolution and a multispectral scene (red, blue, and green) at 4-m resolution. The 1-m resolution imagery is said to be the highest resolution imagery available to the commercial market and is of sufficient detail to discriminate between a car and a truck. The resolution is not high enough to resolve individual people.

The satellite is in a 680-km, sun-synchronous orbit and can obtain an 11-km swath of data at 1-m resolution. The optical telescope has a focal length of 10m and uses five mirrors to image at the focal plane. Combining these mirrors reduces the telescope length to about 2m. The panchromatic imager is composed of 13,500 detectors, where each detector is 12 μm wide. The multispectral detector is a quad-linear detector array with a total of 3375 detectors, where each detector is 48 μm in size. The imagery data is collected at 11 bits per pixel and is compressed to 2.6 bits per pixel before transmission. The panchromatic response is from 0.45 to 0.9 μm, and the spectral responses are as follows: blue (0.45–0.52 μm), green (0.52–0.60 μm), red (0.63–0.69 μm), and NIR (0.76–0.90 μm).

The metric positional accuracy is 12m in the horizontal direction and 10m in the vertical direction, with no ground control, and within 2m with ground control. The orbit time is 98 minutes and the revisit time for the 1-m resolution imagery is 2.9 days. An example image of Beijing is shown in Figure 2.27. Space Imaging, Inc. is marketing imagery from IKONOS under the CARTERRA brand name. They also provide imagery from Indian satellites, European Space Agency (ERS) assets, RADARSAT, and the Japanese imaging system (JERS).

2.7.6 Russian Satellite Imagery

There are a wide variety of Russian satellite data available to the commercial market. One of the distributors of this data is ten-to-ten at www.tentoten.co.uk. This data corresponds to a number of different archive years, footprints, and resolution. Table 2.3 summarizes some of the sensor imagery that is available, including the acquisition sensor. The DD5 data is actually derived from a number of Russian military sources, but is available for sale. Geo-referenced data is sold by the square kilometer.

Figure 2.27 IKONOS image of Beijing, China (courtesy of Space Imaging, Inc.).

Data is also available from SPIN-2 for both the KVR-1000 and the TK 350 [23]. The last mission flown was in 1998. The 1998 mission covered much of the southeastern United States. The TK-350 and the KVR-1000 are film return systems, although their data is digitized for distribution. The two systems are complementary and acquire overlapping coverage. The TK-350 is a 10-m resolution topographic camera that acquires imagery at a scale of

Table 2.3
Commercially Available Russian Imagery

Camera	Footprint	Bands (μm)	Resolution (meters)	Archive years
MK-4	150 × 150 km	0.635–0.690	8	1988–1995
		0.810–0.900	15	
		0.515–0.565	15	
KFA-1000	80 × 80 km	Pan	5	1974–1993
KFA-3000	21 × 21 km	Pan	2	1978–1993
KVR-1000	40 × 40 km	Pan	2	1983–1998
TK-350	200 × 300 km	Pan	10	1983–1998
DD5	By km²	Pan	2	1990+

1:660,000 with 80% overlap between frames. This provides stereo coverage for use in orthorectification of the KVR-1000 imagery.

The KVR-1000 imagery is acquired at a scale of 1:220,000. The imagery is digitized with 1.58-m pixel spacing. The information content may be somewhat better than implied by the stated 2-m resolution. The spectral bandpass of the system is 510–760 nm.

2.7.7 Other Satellite Systems

A rather wide variety of other past, current, and planned satellite systems have or will have imagery available for sale. The SPOT system (SPOT Image) is a family of MSI satellites that began in 1986 [24]. The SPOT program was developed in France (CNES) with support from Belgium and Sweden. SPOT flies in a circular quasi-polar orbit. SPOT 2 and 4 are currently operational (January 2000). SPOT 2 is a three-band MS system, launched in 1990, operating in the visible spectrum; SPOT 4 (launched in 1998) extends to the NIR. Both also have a pan band. SPOT 2 and 4 have a 20-m GSD for the MSI and a 10-m GSD for the pan. SPOT 5, to be launched in 2001, will be similar to SPOT 4 but at finer GSD. Table 2.4 summarizes the relevant characteristics. Information on SPOT is available at www.spotimage.fr/home/present/welcome.htm.

Table 2.4
SPOT Characteristics

System	Band(s)	Wavelength (μm)	GSD
SPOT 2	MS	0.50–0.59	20m
		0.61–0.68	
		0.79–0.89	
	Pan	0.51–0.73	10m
SPOT4	MS	0.50–0.59	20m
		0.61–0.68	
		0.79–0.89	
	NIR	1.58–1.75	20m
	Pan	0.61–0.68	10m
SPOT5	MS	0.50–0.59	10m
		0.61–0.68	
		0.79–0.89	
	NIR	1.58–1.75	20m
	Pan	0.51–0.73	2.5/5m

CORONA is a U.S. intelligence satellite flown between 1962 and 1972. The final version of the system (KH-4B) acquired photographic imagery at a scale of 1:247,500 and an estimated ground resolution of 6 ft [25]. An estimated 18 million square nautical miles of domestic coverage and 379 million square nautical miles of foreign coverage are available. CORONA imagery is maintained by the USGS and can be searched at www.edc.usgs.gov/Webglis/glisbin/search.pl?DISP.

Space Imaging launched Quickbird in the fall of 2000 but a successful orbit was not achieved. Another launch is planned for 2001. It has characteristics very similar to IKONOS [26]. Orbital Sciences has proposed launch of an MS system with 1-m pan and 4-m MS capability (Orbimage 3) as well as a system with the same capabilities plus a 200-channel HS system with 8-m GSD (Orbimage 4) [26]. Both India and Japan have proposed several MS systems; Japan is also proposing to launch a series of 10-m SAR systems [26].

References

[1]   Driggers, R., P. Cox, and T. Edwards, *Introduction to Infrared and Electro-Optical Systems*, Norwood, MA: Artech House, 1999, pp. 1–12.

[2]   Holst, G., *Electro-Optical Imaging System Performance*, Winter Park, FL: JCD Publishing, 1995, p. 3.

[3]   Lloyd, J., *Thermal Imaging Systems*, New York, NY: Plenum Press, 1975, pp. 26–29.

[4]   Anderson, R., et al., *Military Utility of Multispectral and Hyperspectral Sensors*, Ann Arbor, MI: Environmental Research Institute of Michigan, 1994, p. 2-1.

[5]   Wolfe, W., and G. Zissis, *The Infrared Handbook*, Ann Arbor, MI: Environmental Research Institute of Michigan, 1978, pp. 1-5–1-17.

[6]   Pinson, L., *Electro-Optics*, New York, NY: Wiley, 1985, p. 168.

[7]   Kopeika, N., *A System Engineering Approach to Imaging*, Bellingham, WA: SPIE, 1998, p. 331.

[8]   Burle Industries, *Electro-Optics Handbook*, Lancaster, PA: Burle Industries, 1974, p. 180.

[9]   Society of Motion Picture and Television Engineers SMPTE 170M, *Composite Analog Video Signal, NTSC for Studio Application*, White Plains, NY, 1994.

[10] Dereniak, E., and G. Boreman, *Infrared Detectors and Systems*, New York, NY: Wiley, 1996, p. 200.

[11] Shumaker, D., J. Wood, and C. Thacker, *Infrared Imaging Systems Analysis*, Environmental Research Institute of Michigan: Ann Arbor, MI, 1988, p. 10-1.

[12] Leachtenauer, J., et al., "General Image Quality Equations: GIQE," *Applied Optics*, Vol. 36, No. 32, November 1997, pp. 8322–8328.

[13] Schlessinger, M., *Infrared Technology Fundamentals*, New York, NY: Marcell Dekker, 1995, p. 429.

[14] Hovanessian, S., *Introduction to Sensor Systems*, Norwood, MA: Artech House, 1988, pp. 125–148.

[15] Zmuda, Z., and E. Toughlian, *Photonic Aspects of Modern Radar*, Norwood, MA: Artech House, 1994.

[16] Mileshosky, B., *What Is Synthetic Aperture Radar?* Sandia National Laboratories: www.sandia.gov/radar/whatis, 1999.

[17] Freeman, T., *What Is Imaging Radar?* Jet Propulsion Laboratory: www.southport. jpl.nasa.gov/desc.imagingradarv3, 1999.

[18] Information courtesy of Aeronautical Systems, Inc.

[19] Information courtesy of Raytheon.

[20] Information courtesy of NASA Goddard Space Flight Center.

[21] Information courtesy of the Canadian Space Agency.

[22] Information courtesy of Space Imaging, Inc.

[23] Information courtesy of SPIN-2 Corporation.

[24] Information courtesy of SPOT Image.

[25] McDonald, R. A., *CORONA: Between the Sun and the Earth: The First NRO Reconnaissance Eye in Space*, Bethesda, MD: American Society of Remote Sensing and Photogrammetry, 1997

[26] Mitretek Systems, *Land Satellite Information in the Next Decade*, Mitretek Systems, Vienna, VA, 1995

3

Historical Review

In previous chapters, we have discussed S&R systems and introduced the modeling of such systems. Here, we provide a brief history of the development of S&R systems, as well as an historical overview of the accompanying research on information extraction performance. We begin with the pre-1950 timeframe and continue in two periods to the present. Both military and civilian applications are covered, although the latter less thoroughly than the former. The emphasis is on performance prediction and modeling. As is sometimes the case, the historical record is characterized by multiple trails that diverge and converge. We characterize each period with a timeline relating key world events, key S&R developments, and performance prediction-related research. We also correlate major S&R hardware developments with 20th century conflicts.

3.1 Pre-1950

Figure 3.1 diagrams some of the key events of the pre-1950 period. As early as 1912, small cameras were attached to rockets to acquire photos of the Earth. Although military reconnaissance and surveillance are as old as warfare, the use of imaging devices by the United States began in earnest during World War I.[1] Aerial photography was used by the Army to perform reconnaissance

---

1. Although photographs had been taken from balloons prior to the Civil War, cameras were impracticably cumbersome. Civil War balloon reconnaissance used maps and sketches to record observations.

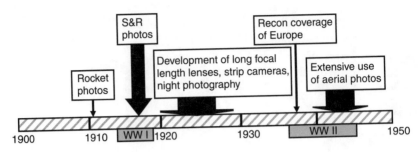

Figure 3.1 S&R timeline: 1900–1950.

from aircraft using vertical, oblique, and stereo photography [1]. The mode of operation paralleled current practice in that specialists (photointerpreters)[2] other than the photographer extracted information from the photos.

At the outset of the war, the United States was not well trained or equipped. U.S. equipment and training was provided by the French and British or relied on French and British expertise. The French and British had been fighting for more than 2 years before the United States entered the conflict. Aerial observation during World War I had evolved along several fronts including balloon observation, aircraft observation, balloon and aircraft artillery spotting, and aerial photography. Aerial photography was typically performed from two-seat biplanes with the observer seated behind the wings using a hand-held camera.

After a fact-finding mission in 1917, the United States began setting up schools to train the various specialists needed to conduct aerial observation and reconnaissance. This included flight training, gunnery, communications, and aerial photography. Schools were also established for support functions such as maintenance, supply, engineering, photo processing, and photo interpretation. The first training school for ground personnel opened at Wilbur Wright Field in September 1917. Wright Field subsequently became Wright-Patterson Air Force Base, a sponsor of many research programs relating to aerial reconnaissance and surveillance.

Because of a lack of qualified instructors and the lengthy period of time required to develop training facilities, as well as the lack of suitable aircraft, training in the United States was at the basic level, with advanced training

---

2. The terms "photo interpreter" and "photointerpreter" were used until sometime in the 1960s. They were replaced by the term "image interpreter" to reflect the growing importance of nonphotographic imagery. Currently, the term "imagery analyst" is used in reference to one who extracts information from aerial imagery.

taking place in France. In France, pilots were trained to fly current aircraft and observers were trained in the data collection, communication, and gunnery required to survive in the war. As they were trained, they assumed the same roles as their French and British counterparts. They performed aerial observation and artillery spotting and acquired and analyzed aerial photos for purposes of military planning.

Following World War I, development of aerial photography continued. One of the pioneers was then Lt. George W. Goddard. Goddard was trained at the first photo interpretation school established at Cornell University during WWI and retired as an Air Force General [2]. He pioneered several equipment developments following WWI, which included night photography using flash (an exploding towed glider), long focal length lenses, and the strip camera for low-altitude reconnaissance. Goddard also pioneered in color photography, high-altitude reconnaissance, and stereo photography. In 1927, he demonstrated near-real-time imaging by taking an aerial photo of Fort Leavenworth, processing it in the air, and transmitting the result by telegraph to New York. Essentially this same technique was proposed for the first U.S. reconnaissance satellite program some 30 years later. Goddard continued his work throughout WWII and beyond with the continued development of high-altitude photography as well as lightweight systems suitable for use in space.

In addition to the work of Goddard, the period leading up to WWII saw continued development in the understanding of S&R imaging. Even prior to 1900, Hurter and Driffield developed an analytic approach to photography [3]. The Fourier domain approach to the analysis of the photographic process was initially developed by Duffieux and Freiser in the 1930s [4]. As early as 1934, problems of sampled imagery were considered in the context of television by Mertz and Gray [5]. The poor initial quality of photographs presented on TV led to considerable research on sampling effects [6].

During WWI, S&R missions were overt and no attempt was made to hide the purpose of the mission. Perhaps one of the first covert S&R missions was carried out by Sidney Cotton [7] shortly before the start of WWII. Recognizing the need for intelligence, Cotton flew his private plane over Europe and photographed a variety of strategic targets. At the start of WWII, he turned over his photos to the Royal Air Force.

World War II saw extensive use of aerial photography for reconnaissance and surveillance. A 1954 U.S. Joint Service manual [8] shows some 38 camera models ranging in focal length from 6 to 40 inches. The 6-inch focal length tri-metragon camera was a popular reconnaissance system (Figure 3.2). The three cameras were mounted side by side with overlapping fields

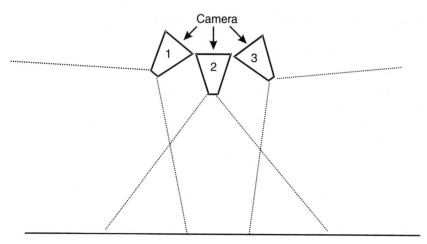

Figure 3.2 Tri-metragon camera field of view.

of view. This system provided, in theory, a 193-degree field of view (FOV). The useful FOV was much smaller because image scale[3] decreased and haze increased as the angle increased. Tri-metragon cameras used both between-the-lens shutters and focal plane shutters.

The use of resolution targets to test cameras was common in WWII. Wright Field, under the leadership of then Major George Goddard, became a center for continued camera development. One of the key developments was a 40-inch focal length camera that could automatically adjust to pressure and temperature variations at high altitude [9]. The concept of strategic overflight reconnaissance was developed at Wright Field as a counter to the threat of atomic warfare. It was recognized that advance warning of intentions, as opposed to simply targeting information, was a necessity.

Although photo interpretation played a key role in WWII, as documented by several authors [10,11], research focused on camera and film development. In the field of photointerpretation, dichotomous keys (successively branching decision trees) were developed as aids to photointerpretation. The emphasis in this period was on training as opposed to performance measurement and prediction.

Information theory was developed in the 1940s for communications applications [12]. By 1950, information theory had been applied to an understanding of the photographic process, and Rose had demonstrated the role of noise in image quality [13].

---

3. Image scale and other image quality metrics are defined in detail in Chapter 6.

Despite the development of much of the theoretical background for current modeling of imaging systems, little application was made in the actual surveillance and reconnaissance process. The tools to apply theory to practical applications did not yet exist. Throughout the 1940s [8] image scale was used to define tasking requirements for aerial photography.

3.2 1950–1970

The period from 1950 to 1970 saw an explosion in research related directly to S&R performance prediction and modeling. The first satellite S&R systems, as well as IR and radar S&R systems, were developed in this period (Figure 3.3).

The Korean War, which began in 1950, saw the development of new camera systems for low level reconnaissance. The continuing Cold War also spawned Allied concerns over Soviet bomber and missile capabilities. At this point in time, countries maintained sovereignty over the airspace above them. Thus, strategic reconnaissance was illegal. Initial attempts were made to gather strategic reconnaissance information by standoff missions flown along the coast of the northern USSR, but as standoff distances were increased by treaty, success was limited. With the detonation of the first Soviet atomic bomb, the need for intelligence began to outweigh the language of the treaties. This led to the first overflights in 1951 by the British using twin-engine USAF jet aircraft (RB45-C). This was followed by RB-47 flights, the last one of which occurred in 1954, the aircraft sustaining damage from fighter aircraft [9]. In an attempt to develop an aircraft that would fly at an altitude safe from hostile fire, the U-2 was designed by Lockheed in 1954. The first test flight took place in 1955.

Recognizing the need to avoid surprise attack, President Eisenhower proposed an "Open Skies" policy at the Geneva Summit Conference in July

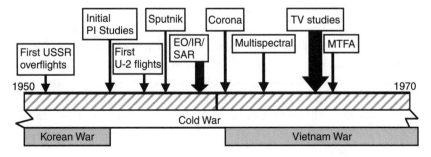

Figure 3.3 S&R timeline: 1959–1970.

1955. A joint program of aerial photographic coverage was proposed. Premier Khrushchev rejected the proposal [9]. Following this rejection, Eisenhower announced a plan to build and launch a series of small, unmanned satellites to support the International Geophysical Year. The underlying intent was to establish the principle of freedom of space, which, in turn, set the stage for the launch of the first reconnaissance satellite in 1959.

Continued concerns over Soviet capabilities led to U-2 overflights beginning in 1956. The Soviets were immediately aware of the overflights and made preparations to defend themselves. The first flight was followed by several successful overflights. On the one hand, the Soviets developed increasingly high-altitude surface-to-air missiles, while on the other, the U-2 program attempted to reduce the radar signature of the U-2 [9]. The Soviets finally succeeded; Gary Powers was shot down in 1960 [14].

The Soviets launched Sputnik in October 1957, causing considerable U.S. concern. The only U.S. reconnaissance satellite program at the time was the Air Force WS 117L program. The WS 117L was designed by RAND Corporation[4] to carry a Kodak camera with a ground resolution of 100 ft and to scan the photos and transmit them to a ground station. An estimated five or six photos per day was the transmission bandwidth-limited capacity of the system [9].

As an alternative to the W117L design, some RAND Corporation scientists, including Amrom Katz, advocated the use of a film return system as a means of increasing the number of images per day. This proposal ultimately came to fruition as the Corona Program in early 1958 under the leadership of Richard Bissell. Itek was the camera developer. The first Corona launch was in June 1959; but this mission did not return imagery. In fact, the first eight flights were unsuccessful. The ninth launch in 1960 was the first successful mission. Over the life of the program (1959–1972), 144 satellites were launched, of which 100 were successful. Imagery ground resolution improved from 40 to 6 ft over the life of the program [15]. Figure 3.4 compares Corona imagery from 1965 (25-ft resolution) with imagery of the same area acquired in 1972 (6-ft resolution).

The Soviets also developed satellite reconnaissance systems [16] with the same motivation as the United States. The Soviets were concerned about U.S. military threats. The Zenit program began in the early 1960s as an outgrowth of earlier reconnaissance satellite development efforts. The Zenit-2 system became operational in 1964 after some 13 launches over the period 1961–1963. Although ground resolution was stated at 10–15m, the

---

4. The RAND Corporation was an Air Force-funded and -directed research laboratory.

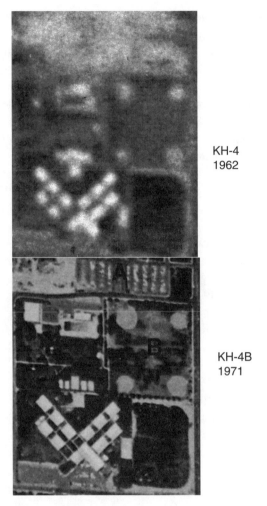

KH-4
1962

KH-4B
1971

Figure 3.4 Comparison of KH-4 and KH-4B imagery.

system had a 1000-mm focal length lens and flew at an altitude of 200 km. This would provide a scale of 1/200,000. Assuming a film resolution of 20 lp/mm, this would provide a ground resolution on the order of 10m. Because commercially available film resolutions on the order of 50 lp/mm were available at the time, a ground resolution of 4m would not be surprising.[5] The Zenit-4 follow-on system reportedly first flew in 1963 and may have become

5. The current KVR 1000 flies at 200 km and has a 1000-mm focal length. The film is digitized with a pixel spacing of 1.58m and the estimated GSD is between 1 and 1.5 m.

operational in 1964. It had an estimated focal length of 3000 mm, providing an estimated ground resolution of 1m.

By the early 1960s, IR and noncoherent side-looking radar (SLR) systems were in use in the military reconnaissance community. The first coherent synthetic aperture radar (SAR) system was developed by the Environmental Research Institute of Michigan (ERIM) in 1958 [17]. The 1967 *Image Interpretation Handbook* [18] described and showed imagery from IR and SLR systems; it also described coherent radar systems (SAR). A 1969 *Reconnaissance Reference Manual* listed three radar and three IR systems in operational use [19].

The Vietnam War (from October 1961 to January 1973) was often fought in heavy jungle and so was a challenge for conventional reconnaissance techniques. IR and radar offered additional potential. In addition to IR and radar, panoramic cameras were also introduced in this era. Panoramic cameras provided wide-angle coverage at high speed. Camera types of this era are diagrammed in Figure 3.5.

In the late 1950s, EO systems began to develop as alternatives to photographic systems. Early systems used vidicon and orthicon tubes operating in a manner similar to home video cassette recorders (VCRs). The first multi-

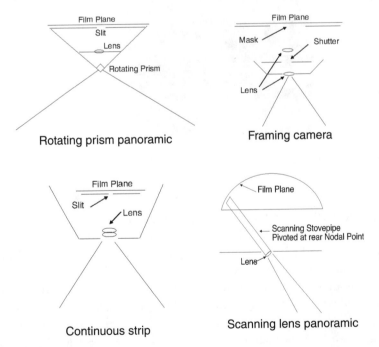

Figure 3.5 Camera types.

spectral scanner was developed in 1962 by ERIM [17]. In the 1960s, work began on development of a civilian satellite remote sensing system. The U.S. Geological Society led the initial effort. Once the utility of a multispectral capability was recognized, research focused on defining the optimum spectral bands for implementation. The growing importance of nonphotographic systems was suggested by an article published in 1961 entitled "Aerial Photographs Are Obsolete" [20]. The article looks to the future with a discussion of EO, IR, radar, and space systems.

Current EO systems collect digital data and transmit it to the ground, where it is displayed on film or cathode ray tubes (CRT). An alternative to EO systems that was developed during the 1960s was monobath processing, followed by scanning for subsequent storage or transmission [21]. In monobath processing, the film developer and fixer are combined and the process is insensitive to development time. Polaroid film is perhaps the most common example. Monobath processing made in-flight film development possible. Subsequent scanning of the film made electronic transmission possible. This was the method proposed for the WS 117L system [9].

Research related to photo or image interpretation largely began in university laboratories funded by the government. The Boston University Optical Research Laboratory was formed in 1946 after Harvard shut down the Harvard Optical Laboratory to return to peacetime pursuits. The BU lab subsequently formed the nucleus of the Itek corporation, which was instrumental in the development of the Corona photographic systems [9]. In 1954, in a preface to a Boston University study on selection tests for photointerpreters, the authors wrote, "Although other components in the reconnaissance system have often been neglected, none have been neglected quite as much as the photo-interpreter" [22]. Boston University performed several studies dealing with recognition performance. Similarly, the Mapping and Charting Research Laboratory of Ohio State University was funded to perform research on "Human Aspects of Photo Interpretation." The Ohio State work dealt extensively with the process of search [23]. Figure 3.6 shows a pattern of eye movements in searching an aerial image. The University of Michigan Willow Run laboratories, which began in the 1940s and subsequently became ERIM, focused on the development of SAR, IR, and multispectral sensor systems.

Research in the 1950s, with a few notable exceptions, tended to focus on psychophysical behavior of interpreters. Ohio State used a device that tracked eye movement and fixations in performing search. Boston University looked at the effect of factors such as visual aspect angle, size, and shape on target detection and recognition. Blackwell [24] established detection thresholds for small point sources of contrast. Other investigators looked at

Figure 3.6 Eye movement search pattern (first 50 fixations).

visual search behavior [24–26]. A few studies were also performed on tests to select interpreters on the basis of visual and cognitive abilities [27,28].

Johnson developed recognition guidelines for the design of image intensifiers based on the number of TV lines subtending the target [29]. This study formed the basis for several subsequent performance modeling efforts. Toward the end of the decade, effort appeared to shift toward measuring the actual performance of photointerpreters [30,31].

In the civil sector, research was largely devoted to developing or demonstrating information extraction techniques. Applications included forestry, agriculture, geology, land use management, and mapping. This research was a continuation of work begun in the 1930s, and the same approach extends to the present.

Between the 1950s and 1960s, tasking of S&R systems switched from the use of "scale" to the use of resolution as a predictor of performance [18]. The proliferation of camera systems with different resolution capabilities ruled out image scale as a good indicator of information potential. Despite the foundations laid in earlier eras, there was relatively little recognition of the roles of decision theory or the Fourier domain in much of the research on photo and image interpretation performed during this era.

Applied research in the 1960s was heavily funded by a small number of government laboratories, including the Air Force Aerospace Medical Research Laboratory (AMRL), Rome Air Development Center (RADC), the U.S. Army's Behavioral Science Research Laboratory (BeSRL)[6] and Human Engineering Laboratory, and the U.S. Navy's Naval Weapons Center and Naval Ordnance Test Center at China Lake, California. Research was also carried out by some of the major aerospace contractors.

BeSRL research focused on methods of improving the performance of photointerpreters; a battery of tests for PI selection was developed [32]. Several studies investigated the use of teaming arrangements to improve performance or reduce time [33,34]. It was shown that training interpreters on objects that might cause false alarms reduced false alarms. An image quality catalog was developed as a means of categorizing image quality [35]. Virtually all of the work at BeSRL during this time frame dealt with aerial photography.

Research at the Naval Weapons Center centered on television [36,37]. The emphasis was on defining size and resolution requirements for target identification. This work continued from the early 1960s into the 1970s.

Research funded by the Air Force was somewhat more eclectic. Of some 46 studies funded by the Rome Air Development Center or the USAF Avionics and Aerospace Medical Research Lab, 16 dealt with some aspect of radar interpretation and 11 with imagery other than aerial photography. Several studies investigated various aspects of image quality. A review of the literature conducted in 1970, however, concluded, "no single measure of image quality is now available to satisfactorily predict (interpreter) performance" [38]. One of the more interesting aspects of the work performed in this era is the lack of computing power and statistical tools to analyze the data that was collected. Nonlinear relationships were commonly found between image quality and performance (Figure 3.7), but the tools to define a nonlinear regression relationship were not generally available. It is also noteworthy that little of the earlier work on noise analysis and the Fourier domain found its way into the applied research of this period. A major exception was Borough's effort using the modulation transfer function area (MTFA) [39]. In general, all of the work of this era appears to have treated the photo-

---

6. Previously named the U.S. Army Personnel Research Laboratory (USAPRO).

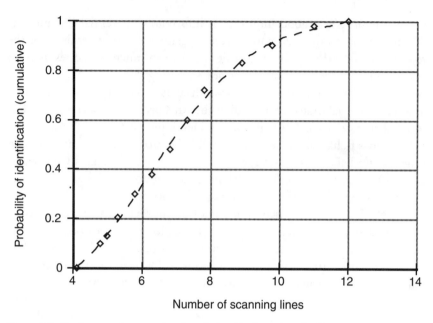

Figure 3.7 Nonlinear performance/quality relationship. (Data from [40].)

interpreter as a black box and did not attempt to model the human perceptual process.

3.3 1970 to Present

Since 1970, there has been a proliferation of civil and commercial satellite imaging systems (Figure 3.8). Beginning with the Earth Resources Technology Satellite (ERTS), at least 18 civil systems have been successfully launched [41]. Initial systems were multispectral EO systems with relatively coarse resolution. Landsat 1 (ERTS-1) had an 80-m resolution (ground sampled distance, GSD), Landsat 7 has 30-m multispectral and 15-m panchromatic resolution [42]. Resolutions improved over this period, and in the 1990s, several commercial ventures were launched. The commercial systems were typically four-band MS systems at 4 or 8 m GSD with a pan capability of 1–2 m GSD. Of the commercial systems to date (September 2000), only the IKONOS system from Space Imaging Inc. has successfully orbited. Orbital radar systems were launched in the 1990s. They included the Russian ALMAZ-1B with ~5-m resolution [41] and RADARSAT with resolution modes ranging from 8 to 50 m [43]. An attempt was made to launch a

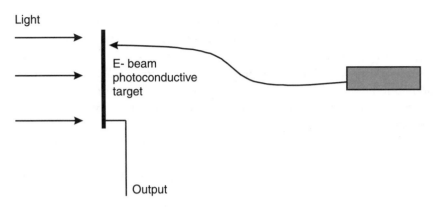

Light

E- beam
photoconductive
target

Output

Figure 3.8 Comparison of image FFTs.

hyperspectral system (named Lewis after Lewis and Clark) in 1998, but the vehicle failed to achieve a stable orbit.

In the military arena, focus continued to shift from photographic to EO systems operating in both the visible and IR spectrum. IR systems improved in both resolution and sensitivity. As an alternative to EO systems, video systems have recently become more popular. Because of their light weight and low cost, they have application in UAV systems. These video systems are generally off-the-shelf camcorders with long focal length optics. UAV systems were also developed with multisensor capability, providing 24-hour, all-weather coverage and including radar, IR, and visible spectrum imagery.

Radar systems improved in resolution. Current systems offer 1-ft resolution and better. Multichannel SARs were developed to explore the effects of frequency diversity [44]. Inverse SAR (ISAR) was developed as a technique to image moving targets such as ships [45]. More recently, interferometric SAR (IFSAR) has been developed to provide three-dimensional data [46].

The proliferation of digital systems has migrated to the consumer market with the marketing of 2-megapixel cameras. Digital video has been developed and is being introduced into the home market. Since the 1970s, multispectral imaging (MSI) has tended to see greater application in the civil as opposed to military market. One of the primary capabilities of MSI is in the analysis of vegetation and soils—with the exception of military trafficability studies, the civil market for MSI is larger. Airborne MSI systems have been developed with 10 to 20 bands; the HYDICE hyperspectral imaging (HSI) system has 206 bands [47]. The large increases in digital data led to a need for data compression (so-called bandwidth compression). Discrete cosine transform (DCT) algorithms were the dominant standard for a number of years; wavelet algorithms have become popular in the last 5 years [48].

The migration to digital systems led to the need for real-time, non-film displays. CRT displays have improved in both resolution and dynamic range over the last 30 years. In the early 1970s, displays had 512×512 pixel addressability with limited dynamic range. Current CRT displays have addressabilities as high as 2048×2560 pixels with a peak luminance as high as 150 fL. Current displays exceed the ability of the human visual system to resolve all of the detail that can be displayed [49]. Flat panel displays such as active matrix liquid crystal displays (AMLCD) can offer even higher levels of detail [50].

The human observer has long been considered the bottleneck in the S&R process. The human is limited in the amount of data that can be processed in a given period of time. Data can be collected at a rate that exceeds the ability of a human to exploit the data at the same rate. Whereas military commanders desire 24-hour, all-weather coverage of the battlefield and UAV systems are becoming capable of providing such coverage, there are simply not enough qualified analysts available to process the data. Maintaining 24-hour coverage with a standard work week and work year requires five people for every staffing level of one. The standard work year is ~1830 hours; 8760 hours are required for 24-hour coverage 365 days a year. Work on replacing the human with machines began in the 1960s and extends to the present [51–53]. As yet, however, this work has not been successful, and the human remains in the loop.

In the arena of modeling S&R performance, the last 30 years built upon work originating in the 1930s through the 1950s. In the Fourier domain, optical power spectrum analyzers were developed [54]. A fast Fourier transform (FFT) of an image was generated optically and the data related to interpreter performance metrics [55]. Figure 3.9 shows two image FFTs generated from an original and a noise-added image. In the FFTs shown in Figure 3.9, power is a function of density (darkness). Note the increased power in the original as opposed to the noise-added image. Theory indicates that, for a given imaging system and set of acquisition conditions, the FFT is independent of scene content. This, in fact, is the case as long as the content is reasonably representative [56]. The same technique, however, has been successfully used to classify image content including land use on aerial photos and fingerprints [57–59]. This work evolved into the digital domain in the 1980s as a tool for quantifying image quality and interpretability [56].

The role of noise in image quality was further developed by Rosell and Willson in the late 1960s and early 1970s [60]. They performed a series of studies, in part based on Johnson's criteria for detection and recognition, that defined the level of signal to noise at the display needed to perform various information tasks. This work formed the foundation for the Minimum Re-

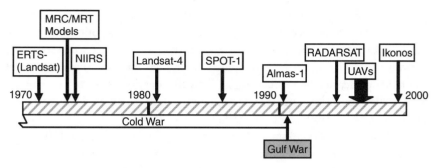

Figure 3.9 S&R timeline: 1970–2000.

solvable Contrast and Temperature models developed by Ratches and described in detail in Chapter 10.

Work on the modulation transfer function approach continued, but limitations were soon recognized [61]. The work at China Lake on video continued into the 1970s [62,63]. Whereas the work in the 1960s had tended to ignore the capabilities and limitations of the human visual system, the 1970s saw increased modeling of the human visual system and incorporation of those models into S&R performance prediction [64,65].

With the advent of digital softcopy[7] systems, work on display quality (mostly CRTs) was carried out. In particular, studies examined ways to optimize the command level/luminance relationship of monitors [66] using empirical and theoretically based approaches. More recently, issues such as addressability and modulation have been studied [49,67].

In the early 1970s, a perceptually based measure called the National Imagery Interpretability Scale (NIIRS) was developed [68]. Using this criteria-based scale, interpreters rated the information content of imagery. NIIRS became a NATO standard in 1976 [69]. The NIIRS is described in detail in Chapter 5 and forms the basis for the general image quality equation (GIQE) described in Chapter 10. Interestingly enough, the same NATO standard carried forward the nonvalidated standards for scale and resolution first published in the 1950s. NIIRS also became the basis for evaluating the performance of physical quality metrics and models including digital power spectrum [59], a just-noticeable difference (JND) model based on a model of the human visual system [65], and the GIQE [70].

---

7. Softcopy refers to a display that can be substantially modified in real time by the observer (using image processing algorithms) as opposed to hardcopy displays such as film and paper prints which are "fixed" in time.

The commercial television (TV) industry developed subjective quality assessment scales for use in evaluating bandwidth compression algorithms and more recently for assessing digital TV data. As was the case with S&R development, effort is now focusing on the development of physical measures that can be used to predict perceptual performance [71]. This effort and related work represent a shift in the conventional military to civilian migration paradigm. Here, commercial standards are migrating to the military arena [72].

Despite the development and advances in SAR, relatively little work appears in the literature relating to SAR quality and interpretability. Although the quality attributes of SAR are understood and documented [73], no extensive modeling efforts have appeared. A somewhat different situation exists with MSI and HSI. Although the physical performance of such systems can be readily quantified, the effect on performance cannot. The potential advantage of such systems is either to increase contrast or to create a unique signature. Most of the work to date appears to have involved determining the ability of a particular system to perform a specific task, as opposed to more general modeling [74,75]. Recently, however, some attempt at modeling HSI has begun at the National Image and Mapping Agency (NIMA) and elsewhere [76,77].

Because it has affected the ability to model S&R performance, some mention should be made of the advances in computing power and general availability. Although mainframe computers were available to large institutions in the early 1960s, they were generally operated in batch mode as opposed to interactively. Statistical analysis packages were generally based on algorithms developed for hand calculation. These limitations tended to restrict the ability to perform in-depth statistical analysis. Although personal computers (PCs) became available in the 1980s, software generally did not migrate from mainframe capabilities to PCs until the 1990s. For what used to be the cost of analyzing data from a single study, an individual can purchase hardware and software to analyze data obtained from the most complex of studies. Whether that capability results in major modeling breakthroughs remains to be seen. These same advances also benefit automated or computer-assisted target recognition. Here, success has yet to be achieved.

3.4 Summary

The last 100 years have seen significant advances in S&R imaging systems. We have moved from hand-held photographic systems to airborne and spaceborne EO and SAR systems with multiwavelength capabilities. The military commanders' desire for 24-hour, all-weather coverage of the battlefield appears to be potentially met with recent UAV systems with both EO and SAR capabilities. Similarly, sensor resolutions have improved to the point where capability often exceeds available bandwidth.

Despite at least 70 years of research effort on image quality, work remains to be done. Effort continues on relating physical quality parameters to information extraction and perceptual rating of quality. In a 1999 summary of image quality modeling, the author writes, "A review of Image Quality Model Building reveals that this field is still in its infancy. There is much work to be done!" [78]. Although this may be the case, useful S&R performance prediction models have been developed and will be discussed in subsequent chapters.

References

[1]   Cobleigh, K. E., *The American Approach to Aerial Reconnaissance and Observation in World War I*, Montgomery, AL: Air Command and Staff College, 1997.

[2]   San Diego Aerospace Museum, "Those We Honor," San Diego, CA: The International Hall of Fame, 1984.

[3]   Hurter, F., and V. C. Driffield, "Photochemical Investigations and a New Method of Determination of the Sensitivity of Photographic Plates," *Journal, Society of Chemical Industrials*, Vol. 9, 1890, pp. 455–469.

[4]   Shaw, R. "A Century of Image Quality," *IS&T Reporter*, Vol. 14, No. 3, 1999, pp. 1–5.

[5]   Mertz, P., and F. Gray, "A Theory of Scanning and Its Relation to the Characteristics of the Transmitted Signal in Telephotography and Television," *Bell System Technical Journal*, XIII, pp. 464–515.

[6]   Biberman, L. M., "Image Quality," in L. M. Biberman (ed.), *Perception of Displayed Information*, New York, NY: Plenum Press, 1973.

[7]   Yost, G., *Spies in the Sky*, New York, NY: Facts on File, 1989.

[8]   Departments of the Army, the Navy, and the Air Force, *Photographic Interpretation Handbook*, TM30-245, Washington, DC: U.S. Government Printing Office, 1954.

[9]   Hall, R. C., "Post War Strategic Reconnaissance and the Genesis of Project Corona," in R. A. McDonald (ed.), *Corona Between the Sun and the Earth: The First NRO Reconnaissance Eye in Space*, Bethesda, MD: American Society for Photogrammetry and Remote Sensing, 1997.

[10] Brugioni, D. A., "Photo Interpretation and Photogrammetry in World War II," *Photogrammetric Engineering and Remote Sensing*, Vol. 50, No. 9, 1984, pp. 1313–1318.

[11] Smith, C. B., *Air Spy: The Story of Photo Intelligence in World War II*, Reprint by American Society of Photogrammetry, Falls Church, VA, 1985.

[12]   Shannon, C. E., and W. Weaver, *The Mathematical Theory of Communication,* Urbana, IL: University of Illinois Press, 1949.

[13]   Rose, A. "The Sensitivity Performance of the Human Eye on an Absolute Scale," *Journal Optical Society of America,* Vol. 38, No. 2, 1948, pp. 196–208.

[14]   Powers, G., "From the U-2 to Corona," in R. A. McDonald (ed.), *Corona Between the Sun and the Earth: The First NRO Reconnaissance Eye in Space,* Bethesda, MD: American Society for Photogrammetry and Remote Sensing, 1997.

[15]   McDonald, R. A., (ed.), *Corona Between the Sun and the Earth: The First NRO Reconnaissance Eye in Space,* Bethesda, MD: American Society for Photogrammetry and Remote Sensing, 1997, Appendix 2.

[16]   Gorin, P. A., "ZENIT: Corona's Soviet Counterpart," in R.A. McDonald (ed.), *Corona Between the Sun and the Earth: The First NRO Reconnaissance Eye in Space,* Bethesda, MD: American Society for Photogrammetry and Remote Sensing, 1997.

[17]   ERIM, *ERIM History,* www.erim-int.com/history_EI.html.

[18]   Naval Reconnaissance and Technical Support Center, *Image Interpretation Handbook,* Volume I, TM30-245, Washington, DC: U.S. Government Printing Office, 1967.

[19]   McDonnell Aircraft Corporation, *Reconnaissance Reference Manual,* Saint Louis, MO: McDonnell Aircraft Corporation, 1969.

[20]   Abrams, T., "Aerial Photographs Are Obsolete," *Photogrammetric Engineering,* Vol. 37, No. 5, pp. 691–694.

[21]   Jensen, N., *Optical and Photographic Reconnaissance Systems,* New York, NY: Wiley, 1968.

[22]   Reyna, L. J., P. Nogee, and S. R. Mayer, *A Study of Two Tests for Discrimination of Proficient Photo-Interpreter Students,* Technical Note 114, Boston, MA: Boston University Optical Research Laboratory, 1954.

[23]   Fry, G. A., and J. M. Enoch, *Second Interim Technical Report: Human Aspects of Photographic Interpretation,* Columbus, OH: Ohio State University Mapping and Charting Research Laboratory, 1956.

[24]   Blackwell, H. R., "Contrast Thresholds of the Human Eye," *Journal of the Optical Society of America,* Vol. 36, No. 11, pp. 624–714.

[25]   Bennett, C. A., et al., *A Study of Image Qualities and Speeded Intrinsic Target Recognition,* Owego, NY: IBM Federal Systems Division, 1963.

[26]   Erickson, C. W., "Partitioning and Saturation of Visual Displays and Efficiency of Visual Search," *J. Appl. Psychol.,* Vol. 39, 1955, pp. 73–77.

[27]   Meyers, J. K., and Miller, R. E., *Validity of Photo Interpreter Predictors for Test and Training Criteria,* WADD-TN-60-45, Lackland Air force Base, TX: Wright Air Development Division, 1960.

[28]   Goldstein, M., A. G. Hahn, and E. L. Chalmers, *A Test of Photo-Interpretation Proficiency,* Laboratory Note 55-2 (Unpublished Draft), Lowry Air Force Base, Denver, CO: Air Force Personnel and Training Research Center, 1955.

[29] Johnson, J., "Analysis of Image Forming Systems," *Proceedings of Image Intensifier Symposium,* Ft. Belvoir, VA, 1959.

[30] Blackwell, H. R., J. Ohmart, and R. Brainard, *Some Psychophysical Factors in Aerial Photo Interpretation,* RADC-TR-61-86, Griffiss AFB, NY: Rome Air Development Center, 1961.

[31] Enoch, J. M., *The Effect on Visual Search of the Degree of Generality of Instructions to the Photointerpreter,* RADC-TN-299, Griffiss AFB, NY: Rome Air Development Center, 1958.

[32] Martinek, H., R. Sadacca, and L. Burke, *Development of a Selection Battery for Army Image Interpreters,* TRN 1143, Washington, DC: U.S. Army Personnel Research Office, 1965.

[33] Bolin, S., R. Sadacca, and H. Martinek, *Team Procedures in Image Interpretation,* TRN 164, Washington, DC: U.S. Army Personnel Research Office, 1965.

[34] Sadacca, R., "The Accuracy and Completeness of Individual and Team Photo Information Extraction," paper presented at the 6th Annual Human Factors Engineering Conference, Ft. Belvoir, VA, October 3–6, 1960.

[35] Brainard, R. W., et al., *Development and Evaluation of a Catalog Technique for Measuring Image Quality,* TRR 1150, Washington, DC: U.S. Army Personnel Research Office, 1966.

[36] Erickson, R. A. and R. E. Main, *Target Acquisition on Television: Preliminary Experiments,* NOTTS TP 4077, China Lake, CA: U.S. Naval Ordnance Test Station, 1966.

[37] Erickson, R. A., and J. C. Hemingway, *Image Identification on Television,* NWC-TP-5025, China Lake, CA: Naval Weapons Center, 1970.

[38] Leachtenauer, J. C. and D. W. Navle, *Image Quality State-of-the-Art Review,* D2-121692-1, Seattle, WA: Boeing Company, 1970.

[39] Borough, H. C., et al., *Quantitative Determination of Image Quality,* D-2-114058-1, Seattle, WA: Boeing Aerospace Co., 1967.

[40] Brainard, R. W., et al., *Resolution Requirements for Identification of Targets in Television Imagery,* NA63H-794, Columbus, OH: North American Aviation, 1965.

[41] Mitretek Systems, *Land Satellite Information in the Next Decade,* McLean, VA: Mitretek Systems, 1995.

[42] Ecosystem Science and Technology Branch, LANDSAT Program Home Page at http:/ /geo.arc.nasa.gov/sge/landsat/landsat.html.

[43] RADARSAT International, *RADARSAT Illuminated—Your Guide to Products and Services,* Version 07195, 1995.

[44] Stuhr, F., R. Jordan, and M. Werner, "SIR-C: An Advanced Radar," in J. M. Kawecki (ed.), *Radar Essentials: Selected Readings,* IEEE, 2000, pp. 401–412.

[45] Voles, R., "Resolving Resolutions: Imaging and Mapping by Modern Radar," in J. M. Kawecki (ed.), *Radar Essentials: Selected Readings,* IEEE, 2000, pp. 167–178.

[46]   Seybold, J. S., and S. J. Bishop, "Three-Dimensional ISAR Imaging Using a Conventional High-Range Resolution Radar," in J. M. Kawecki (ed.), *Radar Essentials: Selected Readings*, IEEE, 2000, pp. 377–382.

[47]   Imagery Resolution and Reporting Standards Committee, *Multispectral Imagery Interpretability Rating Scale*, Newington, VA: Imagery Resolution and Reporting Standards Committee, 1995.

[48]   Mertins, A., "Image Compression via Edge-Based Wavelet Transform," *Optical Engineering*, Vol. 38, No. 6, 1999, pp. 991–1000.

[49]   Leachtenauer, J., A. Biache, and G. Garney, "Effects of Pixel Density on Softcopy Image Interpretability," *Final Program and Proceedings, IS&T PICS Conference*, IS&T, Savannah, Georgia, 1999, pp. 184–188.

[50]   West, J. L., "Liquid Crystal Displays," *Information Display*, Vol. 12, No. 93, pp. 20–23.

[51]   Mundy, J. L., "Observation Events: A Basis for Change Detection," in D. J. Gerson (ed.), *Tools and Techniques for Modeling and Simulation, Proceedings of the 24th APIR Workshop*, SPIE, Vol. 2645, 1995, pp. 89–109.

[52]   Chellappa, R., et al. "Model Supported Exploitation of Synthetic Aperture Radar Images" in D. J. Gerson (ed.), "Tools and Techniques for Modeling and Simulation," *Proceedings of the 24th APIR Workshop*, SPIE, Vol. 2645, 1995, pp. 110–120.

[53]   Hepner, G. F., "Artificial Neural Network Classification Using a Minimal Training Set: Comparison to Conventional Supervised Classification," *Photogrammetric Engineering and Remote Sensing*, Vol. 56, No. 4, 1990, pp. 469–473.

[54]   Kasdan, H. L. "Optical Power Spectrum Sampling and Algorithms," *SPIE Proceedings*, Vol. 117, *Data Extraction and Classification from Film*, 1977.

[55]   Schindler, R. A., *Optical Power Spectrum Analysis of Processed Imagery*, AMRL-TR-79-29, Wright-Patterson Air Force Base, OH: Aerospace Medical Research Laboratory, 1979.

[56]   Nill, N. B., and B. H. Bouzas, "Objective Image Quality Measure Derived from Digital Image Power Spectra," *Optical Engineering*, Vol. 31, No.4, 1992, pp. 813–825.

[57]   Gramenopolous, N., "Automated Thematic Mapping and Change Detection of ERTS-1 Images," *Proceedings, Conference on Management and Utilization of Remote Sensing Data*, Sioux Falls, South Dakota, 1973, pp. 432–446.

[58]   Leachtenauer, J. C., "Optical Power Spectrum Analysis," *Photogrammetric Engineering and Remote Sensing*, Vol. 43, No. 9, 1977, pp. 117–1125.

[59]   Berfanger, D. M., and N. George, "All-digital Ring-wedge Detector Applied to Fingerprint Recognition," *Applied Optics*, Vol. 38, No. 2, pp. 357–369.

[60]   Rosell, F. A., and R. H. Willson, "Recent Psychophysical Experiments and the Display Signal-to-Noise Ratio Concept," in L. M. Biberman (ed.), *Perception of Displayed Information*, New York, NY: Plenum Press, 1973.

[61] Snyder, H. L., "Image Quality and Observer Performance," in L. M. Biberman (ed.), *Perception of Displayed Information*, New York, NY: Plenum Press, 1973.

[62] Lacey, L. A., *Effect of Raster Scan Lines, Image Subtense, and Target Orientation on Aircraft Identification on Television*, Report TP5763, China Lake, CA: Naval Weapons Center, 1975.

[63] Erickson, R., et al., *Resolution of Moving Imagery on Television: Experiment and Application*, NWC TP 5619, China Lake, CA: Naval Weapons Center, 1974.

[64] Carlson, C. R., and R. W. Cohen, *Visibility of Displayed Information: Image Descriptors for Displays*, ONR-CR-120-41, Arlington, VA: Office of Naval Research, 1978.

[65] Lubin, J., *A Methodology for Imaging System Design and Evaluation*, Princeton, NJ: David Sarnoff Research Center, 1995.

[66] Rogers, J. G., and W. L. Carel, *Development of Design Criteria for Sensor Displays*, NR 213-107, Arlington, VA: Office of Naval Research, 1973.

[67] Leachtenauer, J., A. Biache, and G. Garney, "Contrast Modulation—How Much Is Enough?" *Final Program and Proceedings, IS&T PICS Conference*, IS&T, Portland, Oregon, 2000, pp. 130–134.

[68] Leachtenauer, J. C. "National Image Interpretability Rating Scales: Overview and Product Description," *ASPRS/ASCM Annual Convention & Exhibition*, Vol. 1, April 1996, Baltimore, Maryland, pp. 262–271.

[69] North Atlantic Treaty Organization, Annex C, *Minimum Resolved Object Sizes for Imagery Interpretation*, and Annex G., *Imagery Interpretability Rating Scale*, STANAG 3769, MAS (AIR) (76) 256, Brussels Belgium, 1976.

[70] Leachtenauer, J., et al., "General Image Quality Equation: GIQE," *Applied Optics*, Vol. 36, No. 32, November 1997, pp. 8322–8328.

[71] Corriveau, P., and A. Webster, "VQEG Evaluation of Objective Methods of Video Quality Assessment," *SMPTE Journal*, September 1999, pp. 645–648.

[72] DoD/IC/USIGS Video Working Group, *Video Image Standards Profile*, Version 1.5, October 1999.

[73] Mitchel, R. H., *SAR Image Quality Analysis Model*, Ann Arbor, MI: Environmental Research Institute of Michigan, 1974.

[74] Finco, M. V., and G. F. Hepner, "Modeling Agricultural Nonpoint Source Sediment Yield in Imperial Valley, California," *Photogrammetric Engineering and Remote Sensing*, Vol. 64, No. 11, 1998, pp. 1097–1105.

[75] Jensen, J., et al., "Improved Urban Infrastructure Mapping and Forecasting for BellSouth Using Remote Sensing and GIS Technology," *Photogrammetric Engineering and Remote Sensing*, Vol. 60, No. 3, 1994, pp. 339–346.

[76] Martin, L., J. Vrabel, and J. Leachtenauer, "Metrics for Assessment of Hyperspectral Image Quality and Utility," *Proceedings of the International Symposium on Spectral, Sensing Research*, Las Vegas, Nevada, 1999 (published on CD-ROM 11/00).

[77] Martin, L., et al., "Image Quality Evaluation of AOTF and Grating Based Hyperspectral Sensors," *Proceedings of the International Symposium on Spectral, Sensing Research*, Las Vegas, Nevada, 1999 (published on CD-ROM 11/00).

[78] Engeldrum, P. G., "Image Quality Modeling: Where Are We?" *Final Program and Proceedings, IS&T PICS Conference*, IS&T, Savannah, Georgia, 1999, pp. 251–255.

4

Linear Shift Invariant Imaging Systems

Linear shift invariant (LSI) concepts [1–3] are important in the modeling of imaging systems performance. In this chapter, LSI principles are introduced and imaging is described in both the space domain and the frequency domain. Components of imaging systems are then described by their spatial response or point spread function (PSF). The Fourier transform of a component response gives the component modulation transfer function (MTF). The entire system response is the collective response of the components. Finally, sampled imaging systems are linear, but not shift invariant. The concept of sampling is presented using linear system principles.

4.1 Linearity and Shift Invariance

Linearity requires two properties, superposition and scaling. Consider an input scene, $i(x,y)$, and an output image, $o(x,y)$. Given that a linear system is described by $L\{\}$, then

$$o(x,y) = L\{i(x,y)\}. \qquad (4.1)$$

The superposition and scaling properties are satisfied if

$$L\{ai_1(x,y) + bi_2(x,y)\} = aL\{i_1(x,y)\} + bL\{i_2(x,y)\}. \qquad (4.2)$$

Superposition, simply described, is that the image of two scenes (say, a target scene and a background scene) is the sum of the individual scenes imaged

Imaged separately and added   **Imaged together**

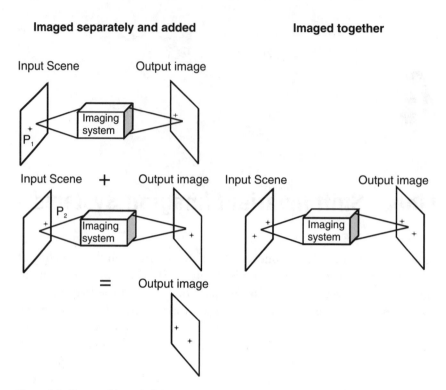

Figure 4.1 Superposition principle.

separately. The simplest example is that of a point source, as shown in Figure 4.1. The left side of the figure shows the case where a single point source is imaged, then a second point source is imaged, and the two results are summed to give an image of the two point sources. The superposition principle states that this sum of point source images is identical to the resultant image if both point sources were included in the input scene.

The second property, scaling, simply states that an increase in input scene brightness increases the image brightness. Doubling a point source brightness would double the output image brightness. The linear systems approach is extremely important with imaging systems because any scene can be represented as a collection of weighted point sources. The output image is the collection of the imaging system responses to the point sources. In continuous (nonsampled) imaging systems, another property is typically assumed —it is the property of shift invariance. Sometimes a shift invariant system is called isoplanatic. Mathematically stated, the response of a shift invariant system to a shifted input, such as a point source, is a shifted output.

$$o(x - x_o, y - y_o) = L\{i(x - x_o, y - y_o)\} \tag{4.3}$$

where x_o and y_o are the coordinates of the point source. It does not matter where the point source is located in the scene. The image of the point source will appear to be the same, only shifted in space. The image of the point source does not change with position.

4.2 The Impulse Function

The impulse function [4], also known as the Dirac delta function, is extremely useful in describing imaging systems because it represents quantities at an infinitesimally small location at a particular position. That is, it can describe a point source of light that is presented to an imaging system. The impulse function is described as a function that has an infinite height, zero width, and an area equal to unity. Mathematically, the impulse function can be described as

$$\delta(x) = \lim_{b \to 0} \frac{1}{|b|} Gaus\left(\frac{x}{b}\right) \tag{4.4}$$

where $Gauss(x/b) = e^{-\pi(x/b)^2}$

If the area of the function remains unity, as b gets smaller, the height of the *Gaus* function increases, as shown in Figure 4.2(a). The mathematical

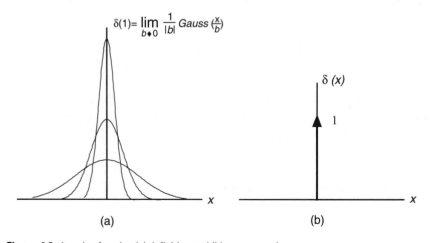

(a) (b)

Figure 4.2 Impulse function (a) definition and (b) representation.

definition given above could have included a number of other functions to replace the *Gaus* function, because the result when *b* goes to zero is the same. In practical systems, *b* can never achieve zero. For example, the impulse function represents a tiny light source input to an optical system in order to obtain the impulse response of the system. While a zero-width light source is not possible, the light source width can be so small that a change in its width has no effect on the output of the system. If this is the case, then the light source can be represented as an impulse function. The graphical representation of the impulse function is shown in Figure 4.2(b). It is an arrow with a height of 1, representing an infinite height, but an area of 1.

There are a few important properties of the impulse response. One of the defining properties of the impulse function is

$$\delta(x - x_o) = 0, \quad x \neq x_o \quad (4.5)$$

so that the only location where the impulse function has a nonzero value is at the location of the impulse function. Another property is the integral property of the impulse function

$$\int_{-\infty}^{\infty} \delta(x - x_o)dx = 1 \quad (4.6)$$

which simply states that the area of an impulse function is 1. The *sifting* property is described as such because the impulse function "sifts" out the value of a function at a particular point:

$$\int_{x_1}^{x_2} f(x)\delta(x - x_o)dx = f(x_o) \quad x_1 < x_o < x_2. \quad (4.7)$$

If an impulse function is placed at some x_o, multiplied by a function, and integrated, the result is the value of the function at the point x_o. Therefore, the delta function has sifted out the value at that point. This property is also important in the process of sampling. Sampling is the process of taking values of a function at discrete, equally spaced locations.

The scaling property of the impulse function is given by

$$\delta\left(\frac{x - x_o}{b}\right) = |b|\delta(x - x_o) \quad (4.8)$$

where this property is intended to account for the area of the impulse function. Recall that the width is infinitesimal and the height is infinite. Also, the impulse function is an even function, so that

$$\delta(x) = \delta(-x). \tag{4.9}$$

These properties are used frequently in the analysis of linear systems. Any function can be decomposed into a combination of impulse functions, where each impulse function has its own impulse response. Then, the response of the system to the input can be determined using superposition.

4.3 The Fourier Series and Fourier Transform

It is necessary to understand Fourier transform principles [5] in the analysis of LSI imaging systems. It is assumed that the reader has some understanding of these principles, but a cursory review is provided in this section.

4.3.1 Fourier Series

Given that a function satisfies certain conditions, the function can be written as a linear combination of other functions that are more fundamental. Typically, these fundamental functions are desired to be an orthogonal set of basis functions. One of these sets of basis functions is that of the complex exponential, $e^{j2\pi\xi x}$, where ξ represents the spatial frequency of the exponential. This basis function is used in the cases of both the Fourier series and the Fourier transform.

If a function is periodic and it satisfies the conditions known as the Dirichlet conditions, the function can be represented by the Fourier series

$$f(x) = \sum_{n=-\infty}^{\infty} c_n e^{j2\pi n \xi_o x} \tag{4.10}$$

where the function is periodic with a period $X = \xi_o^{-1}$. The Dirichlet conditions describe a function within an interval. The function must be single-valued in the integral, must have a finite number of maxima and minima, must have a finite number of discontinuities, and must be absolutely integrable. The complex coefficients in (4.10) are determined by

$$c_n = \frac{1}{X} \int_x^{x+X} f(\alpha) e^{j2\pi n \xi_o \alpha} d\alpha \qquad (4.11)$$

where the coefficients represent the magnitude and phase of the function's harmonic components. The Fourier series can be used to describe a nonperiodic function, but the description is only valid for a finite interval.

4.3.2 Fourier Transform

In order to accurately describe nonperiodic functions, a continuum of fundamental functions is used instead of a set of discrete, harmonically related functions. A nonperiodic function that satisfies the Dirichlet conditions can be represented by

$$f(x) = \int_{-\infty}^{\infty} F(\xi) e^{j2\pi \xi x} d\xi \qquad (4.12)$$

where the weighting function is given by

$$F(\xi) = \int_{-\infty}^{\infty} f(x) e^{-j2\pi \xi x} dx. \qquad (4.13)$$

The weighting function is called the Fourier transform, or the frequency spectrum of $f(x)$. Equations (4.12) and (4.13) are the Fourier transform and the inverse Fourier transform, respectively. Collectively, they are called the Fourier transform pair. In two dimensions, the Fourier transform is

$$F(\xi, \eta) = \int_{-\infty}^{\infty} \int_{-\infty}^{\infty} f(x,y) e^{-j2\pi(x\xi + y\eta)} dx dy \qquad (4.14)$$

and

$$f(x,y) = \int_{-\infty}^{\infty} \int_{-\infty}^{\infty} F(\xi, \eta) e^{j2\pi(x\xi + y\eta)} d\xi d\eta \qquad (4.15)$$

The property of separability means that $f(x, y) = g(x)h(y)$. The spectra are also separable, $f(\xi, \eta) = G(\xi)H(\eta)$. Note that the transform is the product of two one-dimensional transforms. Sometimes it is convenient to use a shorthand notation to denote the process of the Fourier transform. Here, the notation $F(\xi,\eta) = FF\{f(x,y)\}$ and $f(x,y) = FF^{-1}\{F(\xi,\eta)\}$ is used.

Circularly symmetric functions are a special case when it comes to the Fourier transform, so they require special treatment. If $g(r)$ is circularly symmetric and r is the radial variable, then its Fourier transform is

$$G(\rho) = 2\pi \int_0^\infty rg(r)J_o(2\pi\rho r)dr \qquad (4.16)$$

where $\rho = \sqrt{\xi^2 + \eta^2}$ is the radial frequency variable. The shorthand notation is $G(\rho) = BB\{g(r)\}$. The J_o function is the zero-order Bessel function of the first kind. Equation (4.16) is also known as the zero-order Hankel transform. The inverse transform is

$$g(r) = 2\pi \int_0^\infty rG(\rho)J_o(2\pi\rho r)d\rho. \qquad (4.17)$$

Note that the inverse transform is identical to the transform. Also, the transform is circularly symmetric.

4.3.3 Fourier Transform Properties

There are a number of Fourier transform properties that are useful in obtaining the Fourier transform of a complicated function. These properties are given in Table 4.1. Each of the properties assume that $F(\xi) = F\{f(x)\}$,

Table 4.1
Fourier Transform Properties

Space	Frequency
$A_1 f(x) + A_2 h(x)$	$A_1 F(\xi) + A_2 H(\xi)$
$f\left(\dfrac{x}{b}\right)$	$b\lvert F(b\xi)$
$f^*(x)$	$F^*(-\xi)$
$f(x - x_0)$	$F(\xi)e^{-f2\pi x_0 \xi}$
$f(x)*h(x)$	$F(\xi)H(\xi)$
$g\left(\dfrac{r}{b}\right)$	$b^2 G(\rho)$

$H(\xi) = F\{h(x)\}$, and $G(\rho) = BB\{g(r)\}$. Also, A_1, A_2, and b are constants, the superscript * denotes the complex conjugate of the function, and the operator $*$ denotes the convolution operation.

4.3.4 Fourier Transform Pairs

The evaluation of a function's Fourier transform is not usually performed using (4.14) through (4.15). Instead, the transforms of most functions have been previously evaluated and tabulated in terms of Fourier transform pairs. We frequently use these Fourier transform pairs along with the properties described in Section 4.3.3 to determine the transforms of a wide variety of functions. Table 4.2 gives some useful Fourier transform pairs. While the pairs are only listed in one direction, the pairs work in both directions. For example, the table shows that $F(\xi) = F\{\delta(x)\} = 1$. The duality principle states that the Fourier transform from the other side of the pair is $F(\xi) = F\{1\} = \delta(\xi)$.

Table 4.2
Fourier Transform Pairs

Function	Transform
$\delta(x)$	1
$rect(x)$	$Sinc(\xi)$
$tri(x)$	$Sinc^2(\xi)$
$Gaus(x)$	$Gaus(\xi)$
$\cos(2\pi\xi_o x)$	$\frac{1}{2}[\delta(\xi + \xi_o) + \delta(\xi - \xi_o)]$
$\sin(2\pi\xi_o x)$	$\frac{1}{2}[\delta(\xi + \xi_o) - \delta(\xi - \xi_o)]$
$comb(x)$	$comb(\xi)$
$\delta(x,y)$	1
$rect(x,y)$	$sinc(\xi,\eta)$
$tri(x,y)$	$sinc^2(\xi,\eta)$
$Gaus(x,y)$	$Gaus(\xi,\eta)$
$comb(x,y)$	$comb(\xi,\eta)$
$\dfrac{\delta(r)}{\pi r}$	1
$Gaus(r)$	$Gaus(\rho)$

4.4 LSI Imaging System

In an LSI system, the image of a point source does not change with position. If this property is satisfied, the sifting property of the point source or delta function can be used:

$$i(x_o, y_o) = \int_{y_1}^{y_2} \int_{x_1}^{x_2} i(x, y)\delta(x - x_o, y - y_o)dxdy \qquad (4.18)$$

where $x_1 \leq x_o \leq x_2$ and $y_1 \leq y_o \leq y_2$. The delta function, $\delta(x - x_o, y - y_o)$, is nonzero only at x_o, y_o and has an area of unity. The following substitution can be applied to (4.18):

$$i(x,y) = \int_{-\infty}^{\infty} \int_{-\infty}^{\infty} i(\alpha, \beta)\delta(\alpha - x, \beta - y)d\alpha d\beta \qquad (4.19)$$

which states that the entire input scene can be represented as a collection of weighted point sources.

The output of the linear system can then be written using (4.19) as the input, so that

$$o(x,y) = L\left\{ \int_{-\infty}^{\infty} \int_{-\infty}^{\infty} i(\alpha, \beta)\delta(\alpha - x, \beta - y)d\alpha d\beta \right\}. \qquad (4.20)$$

Because the linear operator, $L\{\}$, does not operate on α and β, (4.20) can be rewritten as

$$o(x,y) = \int_{-\infty}^{\infty} \int_{-\infty}^{\infty} i(\alpha, \beta)L\{\delta(\alpha - x, \beta - y)d\alpha d\beta\}. \qquad (4.21)$$

The point source response of the system is called the *impulse response* and is defined as

$$h(x,y) = L\{\delta(x,y)\}, \qquad (4.22)$$

and the output of the system is the convolution of the input scene with the impulse response of the system. That is,

$$o(x,y) = \int_{-\infty}^{\infty} \int_{-\infty}^{\infty} i(\alpha, \beta) h(\alpha - x, \beta - y) d\alpha d\beta = i(x,y)^{**}h(x,y) \quad (4.23)$$

where ** denotes a two-dimensional convolution. The impulse response of the system, $h(x,y)$, is commonly called the point spread function (PSF) of the imaging system. The significance of (4.23) is that the system impulse response is a spatial filter that is convolved with the input scene to obtain an output image. The simplified LSI imaging system is shown in Figure 4.3.

For completeness, we take the *spatial domain* linear systems description and convert it to the *spatial frequency domain*. Spatial filtering can be accomplished in both domains. Given that x and y are spatial coordinates in units of milliradians, the spatial frequency domain has independent variables of ξ and η, in cycles per milliradian. A spatial input or output function is related to its spectrum by the Fourier transform where the inverse Fourier transform converts an image spectrum to a spatial function. From Section 4.3, a spatial function and its spectrum are collectively described as a Fourier transform pair. One of the important properties of the Fourier transform is that the Fourier transform of a convolution results in a product. Therefore, the spatial convolution described in (4.23) results in a spectrum of

$$O(\xi, \eta) = I(\xi, \eta)H(\xi, \eta). \quad (4.24)$$

Here, the output spectrum is related to the input spectrum by the product of the Fourier transform of the system impulse response. Therefore, the Fourier transform of the system impulse response is called the *transfer function* of the system. Multiplication of the input scene spectrum by the transfer function of an imaging system provides the same filtering action as the convolution of the input scene with the imaging system PSF.

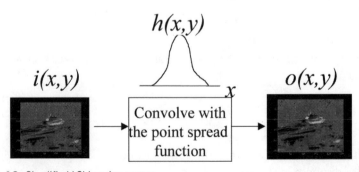

Figure 4.3 Simplified LSI imaging system.

4.5 Imaging in the Space and Frequency Domains

The filtering process is described in (4.23) and (4.24) where one process is in the space domain and the other is in the frequency domain. In space, the output of an LSI system is the input scene convolved with the system impulse response. Take the example given in Figure 4.4. The system shown is a simple imaging system with an input transparency of a four-bar target, an imaging lens, and an output image. Given that the system shown is an LSI system, the output is simply the object convolved with the imaging system impulse response or point spread function. The convolution of the point spread function with the transparency gives a blurred image, as shown. The performance of an imaging system is frequently determined by the response of an imaging system to a four-bar target. The quality of the four-bar target image is evaluated to determine the capabilities of the system.

The equivalent process to the spatial filtering shown in Figure 4.5 is the process through the frequency domain. The Fourier transform of the input function is taken (note that this is a two-dimensional transform). The input spectrum clearly shows the fundamental harmonic of the four-bar target in the horizontal direction. The higher order harmonics are difficult to see in the transform image, because the higher order harmonics have lesser amplitudes than the fundamental. The transform of the point spread function gives the transfer function of the system. In the input spectrum, the transfer function, and the output spectrum, the DC component has been removed so they can be viewed. Otherwise, the magnitude of the DC signal would drown out the rest of the image. Next, the output spectrum is given simply by the input spectrum multiplied by the transfer function. Finally, the output image is given by the inverse transform of the output spectrum. The resulting image is identical to that given by the spatial convolution of the point spread function in the spatial domain.

4.6 Imaging with Components

The system impulse response or point spread function of an imaging system is composed of component impulse responses, as shown in Figure 4.6. Each of the components in the system contributes to the blurring of the scene. In fact, each of the components has an impulse response that can be applied in the same manner as the system impulse response. The blur attributed to a component may be the result of a few different physical effects. For example, the optical blur is a combination of the diffraction and aberration effects of the optical system. The detector blur is a combination of the detector shape

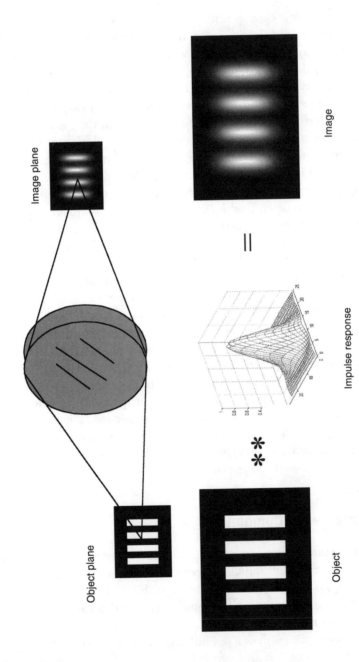

Figure 4.4 Spatial filtering in an optical system.

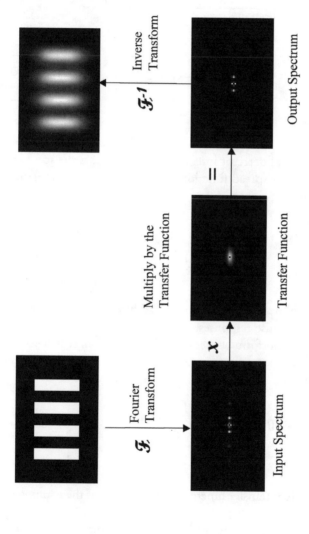

Figure 4.5 Frequency domain filtering in an optical system.

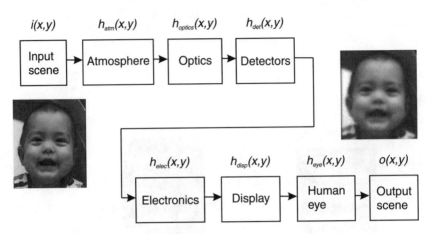

Figure 4.6 System components.

and the finite time of detector integration as it traverses the scene. It can be shown that the point spread function of the system is a combination of the individual impulse responses:

$$h_{system}(x,y) = h_{atm}(x,y) ** h_{optics}(x,y) ** h_{det}(x,y) ** h_{elec}(x,y)$$
$$** h_{display}(x,y) ** h_{eye}(x,y) \qquad (4.25)$$

so that the total blur, or system point spread function, is the convolution of the component impulse responses.

The Fourier transform of the system impulse response is called the transfer function of the system. In fact, each of the component impulse responses given in (4.25) has a component transfer function:

$$H_{system}(\xi,\eta) = H_{atm}(\xi,\eta)H_{optics}(\xi,\eta)H_{det}(\xi,\eta)H_{elec}(\xi,\eta)$$
$$\times H_{display}(\xi,\eta)H_{eye}(\xi,\eta) \qquad (4.26)$$

Note that the system transfer function is the product of the component transfer functions.

There are a large number of imaging spatial filters that are accounted for in the design and/or analysis of imaging system performance. These filters include effects from optics, detectors, electronics, displays, and the human eye. We use (4.25) and (4.26) as our spatial filtering guidelines, where we know that the treatment can be applied in either the spatial or frequency domain. We present the most common of these filters beginning with the optical effects.

4.6.1 Optics

There are two point spread functions (PSFs) that account for the optical effects in an imaging system, diffraction [6] and aberrations [7]. The diffraction PSF accounts for the spreading of the light as it passes an obstruction or an aperture. Diffraction describes the fundamental resolution limit in imaging systems. The diffraction PSF for an incoherent imaging system with a circular aperture of diameter D is

$$h_{diff}(x, y) = \left(\frac{D}{\lambda} \right)^2 somb^2 \left(\frac{Dr}{\lambda} \right) \qquad (4.27)$$

where λ is the average wavelength in the band and $r = \sqrt{x^2 + y^2}$. The *somb* (for sombrero) function is given by Gaskill to be

$$somb(r) = \frac{2 J_1(\pi r)}{\pi r} \qquad (4.28)$$

where J_1 is the first order Bessel function of the first kind. The PSF associated with the optical aberrations is sometimes called the geometric blur. There are many ways to model this blur and there are numerous commercial programs (Code V, Zemax, Oslo, etc.) for calculating geometric blur at different locations on the image. A convenient method, however, is to consider the geometric blur collectively as a Gaussian function

$$h_{gm}(x, y) = \frac{1}{b^2} Gaus \left(\frac{r}{b} \right) \qquad (4.29)$$

where b is the geometric blur in milliradians. The Gaussian function, *Gaus*, is

$$Gaus(r) = e^{-\pi(r)^2} \qquad (4.30)$$

Note that the scaling values in front of the *somb* and the *Gaus* functions are intended to provide a functional area (under the curve) of unity so that no gain is applied to the scene. Examples of the optical impulse responses are given in Figure 4.7, corresponding to a wavelength of 10 micrometers, an optical diameter of 10 centimeters, and a geometric blur of 0.1 milliradians. The overall impulse response of the optics is the combined blur of both the diffraction and aberration effects.

$$h_{optics}(x, y) = h_{diff}(x, y) ** h_{gm}(x, y). \qquad (4.31)$$

The transfer functions corresponding to the above impulse responses are obtained by taking the Fourier transform of the functions given in (4.28) and (4.29). The Fourier transform of the *somb* is given by Gaskill so that the transfer function is

$$H_{diff}(\xi, \eta) = \frac{2}{\pi} \left[\cos^{-1}\left(\frac{\rho\lambda}{D}\right) - \frac{\rho\lambda}{D}\sqrt{1 - \left(\frac{\rho\lambda}{D}\right)^2} \right] \qquad (4.32)$$

where $\rho = \sqrt{\xi^2 + \eta^2}$ and is plotted in cycles per milliradian. The Fourier transform of the *Gaus* function is simply the *Gaus* function, with care taken on the scaling property of the transform. The transfer function corresponding to the aberration effects is

$$H_{gm}(\xi, \eta) = Gaus(b\rho) \qquad (4.33)$$

For the example described above, the transfer functions are also shown in Figure 4.7. Note that the overall optical transfer function is the product of the two functions.

4.6.2 Detectors

The detector spatial filter [8,9] is determined by a number of different effects, including spatial integration, sample-and-hold, crosstalk, and responsivity. The two most common effects are spatial integration and sample-and-hold.

$$h_{det}(x, y) = h_{det\_sp}(x, y) ** h_{det\_sh}(x, y) \qquad (4.34)$$

The other effects can be included but are usually considered negligible unless there is good reason to believe otherwise (e.g., the detector responsivity varies dramatically over the detector).

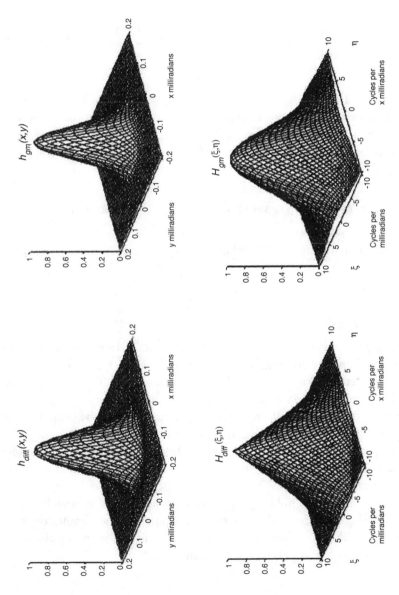

Figure 4.7 Optical PSFs and corresponding transfer functions.

The detector spatial impulse response is due to the spatial integration of the light over the detector. Because most detectors are rectangular, the rectangle function is used as the spatial model of the detector:

$$h_{\text{det\_}sp}(x, y) = \frac{1}{\alpha_x \alpha_y} rect\left(\frac{x}{\alpha_x}, \frac{y}{\alpha_y}\right) = \frac{1}{\alpha_x} rect\left(\frac{x}{\alpha_x}\right) \frac{1}{\alpha_y} rect\left(\frac{y}{\alpha_y}\right) \quad (4.35)$$

where α_x and α_y are the horizontal and vertical detector angular subtenses in milliradians. The detector angular subtense is the detector width (or height) divided by the sensor focal length. The transfer function corresponding to the detector spatial integration is determined by taking the Fourier transform of (4.35):

$$H_{\text{det\_}sp}(\xi, \eta) = \text{sinc}(\alpha_x \xi, \alpha_y \eta) = \text{sinc}(\alpha_x \xi)\,\text{sinc}(\alpha_y \eta) \quad (4.36)$$

where the *sinc* function is defined as

$$\text{sinc}(x) = \frac{\sin(\pi x)}{\pi x}. \quad (4.37)$$

The impulse response and the transfer function for a detector with a 0.1 by 0.1 milliradian detector angular subtense are shown in Figure 4.8.

The detector sample-and-hold function is an integration of the light as the detector scans across the image. This function is not present in staring arrays, but is present in most scanning systems where the output of the integrated signal is sampled. The sampling direction is assumed to be the horizontal or x direction. Usually, the distance, in milliradians, between samples is smaller that the detector angular subtense by a factor, s, called samples per IFOV or samples per DAS.

The sample-and-hold function can be considered a rectangular function in x where the size of the rectangle corresponds to the distance between samples. In the spatial domain y direction, the function is an impulse function. Therefore the impulse response of the sample and hold function is

$$h_{\text{det\_}sb}(x, y) = \frac{s}{\alpha_x} rect\left(\frac{xs}{\alpha_x}\right)\delta(y) \quad (4.38)$$

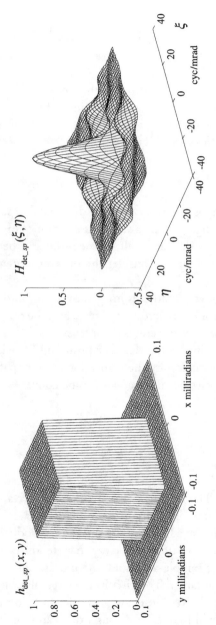

Figure 4.8 Detector spatial impulse response and transfer function.

The Fourier transform of the impulse response gives the transfer function of the sample-and-hold operation

$$H_{\det\_sh}(\zeta, \eta) = \text{sinc}\left(\frac{\alpha_x \zeta}{s}\right) \tag{4.39}$$

Note that the Fourier transform of the impulse function in the y direction is 1. The impulse response and the transfer function for sample-and-hold associated with the detector given in Figure 4.8 with a 2 sample per DAS sample-and-hold are shown in Figure 4.9.

4.6.3 Electronics

Linescanners are composed of a number of different PSFs, one of which is the electronics. Most engineers are aware that electronic amplifiers and filters have transfer functions that are well defined and can be analyzed using Laplace or Fourier transforms. Consider the imaging system shown in Figure 4.10.

In the system shown, a target is imaged by a lens onto an image plane. In this plane, a detector scans across the image and the output of the detector is an electrical signal that represents the intensity variation across the image on the line corresponding to the detector position. The electrical signal is related to the spatial variation by the scan velocity, so a spatial frequency in cycles per meter or cycles per milliradian can be converted to Hertz by the scan velocity:

$$f = \xi v \quad \text{(Hz)} \tag{4.40}$$

where v is the scan velocity in meters per second or milliradians (or the field of view) per second. The circuit is usually an amplifier and a filter but is usually modeled as a filter. The circuit typically drives a display monitor (or a recorder that is then played) where the image is reproduced.

From the previous section, it is clear that the diffraction of the lens, the detector shape, the sample-and-hold, the monitor display spot, and the eye all blur the original target. The circuit, however, also blurs the target, so the circuit has an impulse response and a transfer function that can be converted to space or spatial frequency through the scan velocity.

To consider the electronics, consider an image plane with no diffraction blur, no detector shape blur (i.e., an infinitesimal detector size), and a

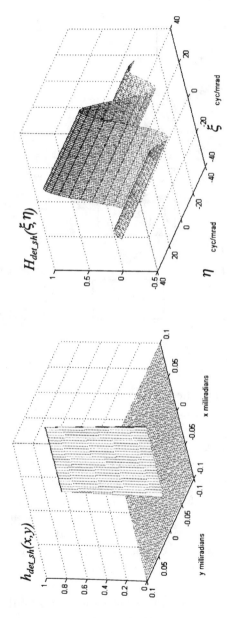

Figure 4.9 Detector sample-and-hold impulse response and transfer function.

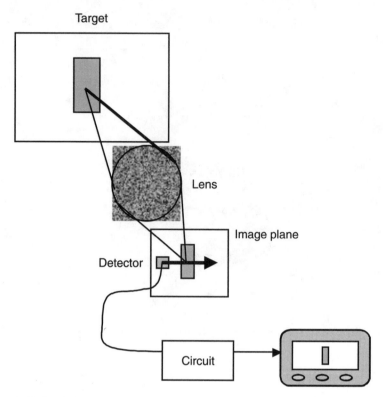

Figure 4.10 Circuit in the imaging system.

perfect display. Therefore, the blur introduced by the circuit is the only blur seen on the display. Electronics can typically be modeled as a low-pass filter, and we use this model as an illustration of electronic blur. Consider the circuit shown in Figure 4.11 along with the circuit input voltage. The input voltage is a scan of the target (with infinitesimal detector size) converted to time by the scan velocity. The target is a rectangular target that produces a quick rise in detector voltage and then a quick drop in detector voltage. The detector voltage is the circuit input signal, where the circuit output voltage is the driving signal of the display. The resistor and capacitor values are set to limit the detector output noise. Both the impulse response and the transfer function of the circuit are desired and the response to the rectangular target is desired.

Most electrical engineers would make short work of this problem using Laplace transforms to determine the response of the circuit to the input voltage. There have been arguments in the past that such a circuit must be

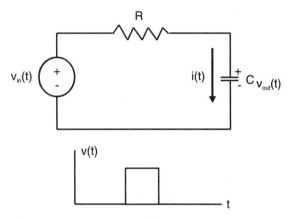

Figure 4.11 Imager circuit and input signal.

analyzed with Laplace transforms because the damping response of the circuit is treated nicely with the complex nature of the Laplace kernel. Fourier transforms, however, work just as well with such a circuit as long as the impulse response is treated as one-sided. That is, a response cannot be seen in time before the stimulation signal occurs on the circuit input. With spatial blurring, impulse functions are typically two-sided, as signals to the left of the spatial origin are well understood. Causality dictates that a circuit cannot respond to an input signal prior to the presence of the input signal (i.e., left of the origin in time).

In the analysis of the low-pass circuit, Kirchoff's voltage equation gives

$$-v_{in}(t) + \frac{1}{C}\int_0^t i(t) + i(t)R = 0 \qquad (4.41)$$

where the initial current is zero. Taking the Fourier transform of the equation gives

$$V_{in}(\omega) + \frac{1}{jwC}I(\omega) + I(\omega)R. \qquad (4.42)$$

The current spectrum in the circuit can be determined from (4.42):

$$I(\omega) = \frac{V_{in}(\omega)}{R + 1/j\omega C}. \qquad (4.43)$$

The output voltage spectrum is simply the current spectrum times the impedance of the capacitor:

$$V_{out}(\omega) = \frac{V_{in}(\omega)}{R + 1/j\omega C} 1/j\omega C. \tag{4.44}$$

The transfer function of the circuit is the ratio of the output spectrum over the input spectrum:

$$H(\omega) = \frac{V_{out}(\omega)}{V_{in}(\omega)} = \frac{1}{1 + j\omega RC} \tag{4.45}$$

The PSF is determined by taking the inverse Fourier transform of the transfer function:

$$h(t) = \frac{1}{RC} e^{-t/RC} step(t) \tag{4.46}$$

This is the function that is convolved with the detector output to give the response of the electronics. Figure 4.12 shows a perfect rectangular image and an output of this image once convolved with the impulse response given by (4.46). Note that the impulse response is one sided (i.e., the step function limits the values of the impulse response to $t \geq 0$).

Figure 4.12 Perfect target and electronics-blurred target.

The amount of electronics blur on the target shown in Figure 4.12 is exaggerated. The electronics of an imaging system usually imparts only a small amount of blur on the image compared to the effects of detector shape and optical blur. In fact, the example given above is one corresponding to scanned detectors. For staring detectors, the electronics blur is usually extremely small, as there are no horizontal or vertical low-pass or high-pass filters in the system. The only blur in the staring array case may be a charge transfer efficiency.

4.6.4 Displays

The finite size and shape of the display spot [10] also correspond to a spatial filtering, or PSF, of the image. Usually, the spot, or element, of a display is either Gaussian in shape like a cathode ray tube or it is rectangular in shape like a flat panel display as shown in Figure 4.13. Liquid crystal devices (LCDs) and light emitting diode (LED) displays are rectangular in shape. They are called flat panel devices [11]. The PSF of the display is simply the size and shape of the display spot. Consider the two displays in Figure 4.13. The display to the left corresponds to the CRT and has a Gaussian display spot. The spot is shown in the lower right corner of the display. This spot is convolved with the scene to obtain the CRT output image, as shown. Flat panel displays, as shown on the right display, have rectangular display elements that can impose display artifacts on the image. This is especially true if the rectangular elements are so large that the edges of the elements are not filtered by the eye.

The finite size and shape of the display spot must be converted from a physical dimension to the sensor angular space. For the Gaussian spot, the

Figure 4.13 (a) CRT display with a Gaussian PSF, and (b) flat panel display with a rectangular PSF.

spot size dimension in centimeters must be converted to an equivalent angular space in the sensor's field-of-view:

$$\sigma_{angle} = \sigma_{cm} \frac{FOV_v}{L_v} \qquad (4.47)$$

where L_v is the length in centimeters of the diplay vertical dimension and FOV_v is the field-of-view of the sensor in milliradians. For the rectangular display element, the height and width of the display element must also be converted to the sensor's angular space. The vertical dimension of the rectangular shape is obtained using (4.47) and the horizontal dimension is similar to the horizontal display length and sensor field-of-view. Once these angular dimensions are obtained, the PSF of the display spot is simply the size and shape of the display element:

$$h_{disp}(x, y) = \frac{1}{\sigma_{angle}^2} Gaus\left(\frac{r}{\sigma_{angle}}\right) \qquad (4.48)$$

for a Gaussian spot, or

$$h_{disp}(x, y) = \frac{1}{W_{angle}H_{angle}} rect\left(\frac{x}{W_{angle}}, \frac{y}{H_{angle}}\right) \qquad (4.49)$$

for a flat panel, where the angular display element shapes in width, W_{angle}, and height, H_{angle}, are given in milliradians. The transfer functions associated with these display spots are determined by taking the Fourier transform of the above PSF equations:

$$h_{disp}(\xi, \eta) = Gaus(\sigma_{angle}\rho) \qquad (4.50)$$

for a Gaussian display, or

$$H_{disp}(\xi, \eta) = sinc(W_{angle}\xi, H_{angle}\eta) \qquad (4.51)$$

for a flat panel display.

4.6.5 Human Vision

The human eye [12] has a PSF that is a combination of three physical components: optics, retina, and tremor. In terms of these components, the PSF is

$$h_{eye}(x, y) = h_{optics}(x, y) ** h_{retina}(x, y) ** h_{tremor}(x, y) \qquad (4.52)$$

The point spread function is rarely ever applied to imagery because if the human views imagery that has been processed with sensor effects, the eye PSF is naturally part of the image blur. We do, however, frequently determine the transfer functions associated with the eye components because we are interested in calculating the performance of the sensor including human vision. Therefore, the transfer function of the eye is [13]

$$H_{eye}(\xi, \eta) = H_{optics}(\xi, \eta) H_{retina}(\xi, \eta) H_{tremor}(\xi, \eta) \qquad (4.53)$$

The transfer function associated with the eye optics is a function of display light level. This is because the pupil diameter changes with light level. The pupil diameter is

$$D_{pupil} = -9.011 + 13.23 \exp\{-\log_{10}(fl)/21.082\} \quad [mm] \qquad (4.54)$$

This equation is valid if one eye is used, as in some targeting applications. If both eyes view the display, the pupil diameter is reduced by 0.5 mm. There are two parameters, *io* and *fo*, that are required for the eye optics transfer function. The first parameter is

$$io = \left(0.7155 + 0.277/\sqrt{D_{pupil}}\right)^2 \qquad (4.55)$$

and the second is

$$fo = \exp\{3.663 - 0.0216 * D^2_{pupil} \log(D_{pupil})\} \qquad (4.56)$$

The eye optics transfer function can be written

$$H_{optics}(\rho) = \exp\{-[43.69(\rho/M)/fo]^{10}\} \qquad (4.57)$$

where ρ is the radial spatial frequency, $\sqrt{\xi^2 + \eta^2}$, in cycles per milliradian. M is the system magnification (angular subtense of the display to the eye divided by the sensor FOV). The retina transfer function is

$$H_{retina}(\rho) = \exp\{-0.375(\rho/M)^{1.21}\} \qquad (4.58)$$

The transfer function of the eye due to tremor is

$$H_{tremor}(\rho) = \exp\{-0.44441(\rho/M)^2\} \qquad (4.59)$$

which completes the eye model.

For an example, let the magnification of the system equal 1. With a pupil diameter of 3.6 mm corresponding to a display brightness of 10 fL at the eye (with one viewing eye), the combined MTF of the eye is shown in Figure 4.14. The *io* and *fo* parameters were 0.742 and 27.2, respectively.

All of the PSFs and transfer functions given in this section are used in the modeling of IR and EO imaging systems. We only covered the more com-

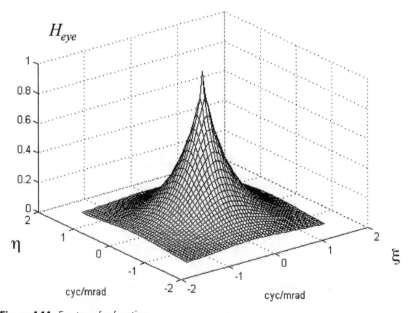

Figure 4.14 Eye transfer function.

mon system components. There are many more that must be considered when they are part of an imaging system.

4.7 Simplifying LSI Imager Analysis to One Dimension

It is common in imaging system analysis to analyze imaging sensors separately in the horizontal and vertical directions. The PSF and the MTF are assumed to be separable in Cartesian coordinates so that the analysis reduces to one dimension and does not require large calculations that include cross-terms. This approach allows straightforward calculations that quickly determine sensor performance.

The separability assumptions are almost never satisfied, so there is some error associated with the sensor performance calculations. While the majority of scientists and engineers still use the approximation of separability, it is important to remember that even in the simplest cases, there is some error associated with the assumption. Separability in Cartesian coordinates requires that

$$f(x, y) = f(x)f(y) \tag{4.60}$$

where separability in polar coordinates requires

$$f(r, \theta) = f(r)f(\theta). \tag{4.61}$$

The optical PSF is a combination of the diffraction spot and the geometric aberrations. Usually, these functions can be characterized by a function that is separable in polar coordinates. The detector PSF is a rectangular shape that is separable in Cartesian coordinates, but is not separable in polar coordinates. The collective PSF of the detector and the optics is a function that is the two-dimensional convolution of the two component PSFs and is not separable in polar or Cartesian coordinates.

An example illustrates typical errors. We take an optical aperture of 16.9 cm at a wavelength of 4.0 micrometers to give a diffraction spot that has a radius (from the center of the spot to the first zero) of 0.024 milliradians. The system is diffraction-limited, so that the geometric or aberration blur is negligible. Next, the DASs in the horizontal and vertical directions are both 0.034 milliradians. To determine the collective PSF of the optics and the detector, the Airy disc and the detector rectangle are convolved. We want to determine the error associated with a one-dimensional approximation of the PSF.

First, we take the two-dimensional PSF as

$$PSF(x, y) = \left(\frac{D}{\lambda}\right)^2 somb\left(\frac{D\sqrt{x^2 + y^2}}{\lambda}\right)$$

$$** \frac{1}{DAS_x DAS_y} rect\left(\frac{x}{DAS_x}, \frac{y}{DAS_y}\right) \quad\quad (4.62)$$

where the ** denotes the two-dimensional convolution. This function is not separable. Next, we take the separable approach where the PSF is estimated as the combination of two separable functions:

$$PSF(x) = \frac{D}{\lambda} somb\left(\frac{Dx}{\lambda}\right) * \frac{1}{DAS_x} rect\left(\frac{x}{DAS_x}\right) \quad\quad (4.63)$$

and

$$PSF(y) = \frac{D}{\lambda} somb\left(\frac{Dy}{\lambda}\right) * \frac{1}{DAS_y} rect\left(\frac{y}{DAS_y}\right) \quad\quad (4.64)$$

Now, the separable PSF is constructed from the two separable functions:

$$PSF(x, y) = \left[\frac{D}{\lambda} somb\left(\frac{Dx}{\lambda}\right) * \frac{1}{DAS_x} rect\left(\frac{x}{DAS_x}\right)\right]$$

$$\times \left[\frac{D}{\lambda} somb\left(\frac{Dy}{\lambda}\right) * \frac{1}{DAS_x} rect\left(\frac{y}{DAS_y}\right)\right] \quad\quad (4.65)$$

The separable function is compared to the nonseparable function in Figure 4.15.

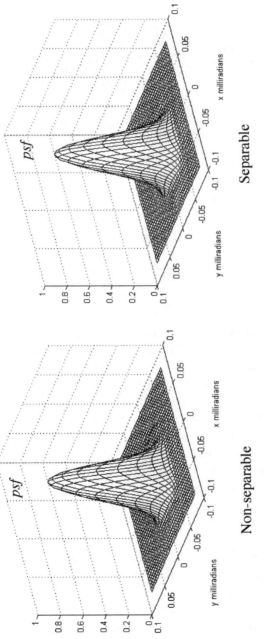

Figure 4.15 Differences in the separable and nonseparable PSFs.

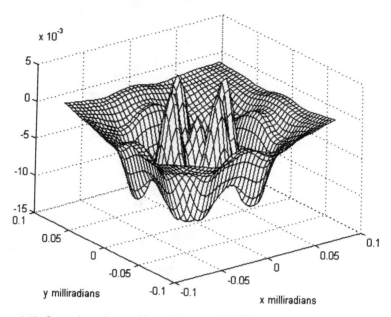

Figure 4.16 Comparison of separable and nonseparable PSFs.

The error can now be determined for the separability assumption (for the optics and detector characteristics stated above). Figure 4.15 shows the separable and nonseparable PSFs and Figure 4.16 shows the error between the PSFs. Note that for the parameters given, the maximum spatial difference is around 1%. An even better comparison is an image that has been blurred by the two PSFs. Consider the images shown in Figure 4.17. The top image is a pristine image that has not been blurred. The image on the lower left has been blurred with the nonseparable diffraction and detector blurring effects. The image on the lower right has been blurred with a separable PSF function. From the imagery, there is no noticeable difference in the blurring effects on the imagery. The process applied to the imagery was a spatial process, so it was the PSF convolved with the pristine image. Keep in mind that the separability issue could have been addressed in the transform domain, where the MTF was applied in one or two dimensions.

In summary, the analysis of imaging systems is usually performed in one dimension and applied to the horizontal and vertical directions. These one-dimensional analyses allow a great simplification to the sensor performance estimates. The separability assumption, however, is not an error-free one. Errors accompany the analyses and, if necessary, can be quantified. Usually, they turn out to be small compared to the other sensor analysis-related errors.

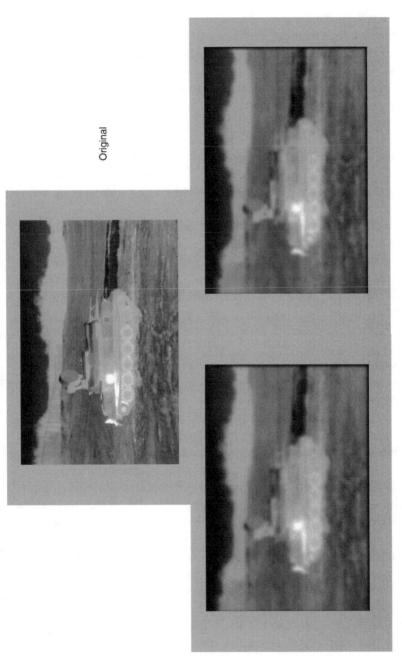

Figure 4.17 Imagery comparison.

4.8 Sampled Imaging Systems

The past few years have seen the development of an extensive number of IR focal plane arrays, and staring array detectors have been available for a long time in EO systems. A common characteristic of IR imagers that use these staring focal planes is that they are spatially undersampled. The performance of undersampled imagers is not well understood. There are numerous efforts currently under way to quantify the performance characteristics of sampled imaging systems.

In general terms, a sampled imaging system [14–20] can be characterized by a three-step process, as shown in Figure 4.18. Note that this process is a much more difficult process than the LSI process given by Figure 4.3. An input image is degraded by a pre-sample blur. This blur includes the combined effects of diffraction and aberrations of the collecting optics, the finite size and shape of the detector, detector scan integration, system vibration, and any other effect that causes image blur prior to the sampling action of the system. The blurred image is then sampled by a finite number of impulse functions that represent the blurred image amplitude at discrete points in space. The third step is the reconstruction, or blurring, of the sampled function to give an output image. The reconstruction blur includes any post-sampling electronic filtering, display spot size, eye blur, and any post sampling blur function.

A well-sampled imaging system is one where the spacing (in milliradians or millimeters) between image samples is small compared to the width of the pre-sample PSF. There are many scientists and engineers who suggest that two samples per detector width (regardless of the optical blur spot size) comprise a well-sampled imaging system. This requirement appears to be an application of the Nyquist sampling requirement that is incorrectly applied to imaging systems. The sampling theorem and the Nyquist sampling requirement are valid only if a perfect reconstruction filter is applied to the sampled signal; this an impossible task with imaging systems. It is interesting to note that the above two-sample-per-detector width usually provides a well-sampled imaging system. A well-sampled imaging system, however, is not a well-designed imaging system. Focal planes are expensive (especially in the infrared) and it is not in the best interest of the designer to provide a larger number of detectors than that of a sufficiently sampled system.

An undersampled imaging system is one where the sample spacing is large compared to the pre-sample blur. In this case, aliasing and spurious responses are inherent to the output imagery and these artifacts can limit the performance of the imaging system. These artifacts are seen as localized deviations in edge locations, shapes, and image intensities. The corruption of

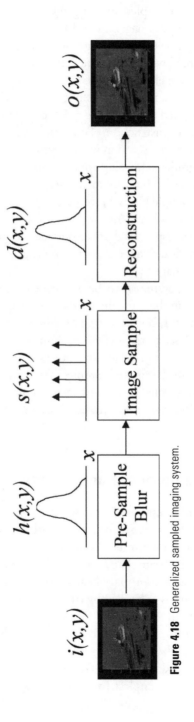

Figure 4.18 Generalized sampled imaging system.

the imagery in an undersampled imaging system is dependent on the frequency content of the input imagery. Band-limited imagery may appear with high quality. Imagery, however, with significant amounts of high-frequency content are accompanied with significant artifacts that limit the overall utility of the imager in these environments.

A well-designed imaging system is not described in such a straightforward way. A well-designed system is the result of a trade between sampling, pre-sample blur, and reconstruction. Too much pre-blur can limit performance, with degradation in resolution while wasting detector count. Too little pre-blur can cause significant aliasing and/or spurious response content that also limits performance with high-frequency scenes. The combination of pre-sample blur and post-sample blur with respect to the sample rate sets the blur versus sampling artifact tradeoff. Also, a poorly designed reconstruction blur can introduce spurious responses that are outside the frequency band of the baseband transfer function. These artifacts are raster and pixelization effects than can ruin output imagery.

Sampled imaging systems can be analyzed in the frequency domain (with Fourier analysis) in the same manner as LSI imaging systems. Sampled imaging systems are linear, but they are not shift-invariant. It is beyond the scope of this book to describe the analysis of sampled imaging systems. There are entire books dedicated to this subject. For example, see [13]. In short, a sampled imaging system that is well sampled can be treated as an LSI imaging system. An undersampled imaging system cannot be treated as an LSI imaging system, and references are provided for the treatment of these systems.

4.9 SAR Impulse Response and Transfer Function

SAR [21] has an impulse response in the same manner as EO and IR imaging systems. The impulse response for a radar without the synthetic aperture integration length can be determined from the antenna length. The real aperture size determines the beam shape, in angle. This beam shape corresponds to the impulse response, and primary beam-broadening is caused by diffraction. When the aperture is enhanced by the synthetic aperture distance, the beam is reduced in width.

In its simplest form, the impulse response of a SAR system is determined by the filtering of the returned signal. Also, because SAR is a coherently illuminated and processed signal, the transfer function can be determined directly by the filter. The simplest form is a bandpass filter of rectangular form:

$$MTF(\xi, \eta) = rect(2\Delta R_x \xi, 2\Delta R_y \eta) \qquad (4.66)$$

where ΔRx and ΔRy are given as the SAR resolution in Chapter 2. The impulse response, or IPR as it is described by SAR engineers, is determined by the Fourier transform of the MTF

$$IPR = 0.25 \, sinc\left(\frac{x}{2\Delta x}, \frac{y}{2\Delta y}\right) \qquad (4.67)$$

Equation (4.71) is an oversimplified approximation of the impulse response and finite impulse response (FIR) filters are used frequently to match the return signal of the radar. In truth, the IPR of the radar is more closely related to the matched filter design [22] of the system that is optimized for both resolution and signal-to-noise. It is beyond the scope of this book to cover the matched filter analysis and design of these systems. Additional information can be found in [23].

4.10 Summary

We described the properties of a linear system—superposition and scaling. With nonsampled imaging systems, the property of shift invariance is assumed.

Imaging systems were described in the spatial domain in terms of their impulse response (the response of the system to a point source of energy). A scene can be considered as a collection of weighted point sources and the output of the system is the convolution of the input scene with the system impulse response. The system response is the result of a combination of individual component responses ranging from the optics through the response of the display and the human visual system.

In the frequency domain, the Fourier transform of the input function is multiplied by the Fourier transform of the system response function. The inverse transform defines the output image.

References

[1]   Goodman, J. *Introduction to Fourier Optics,* San Francisco, CA: McGraw-Hill, 1968, pp. 17–25.

[2] Gaskill, J. *Linear Systems, Fourier Transforms, and Optics,* New York, NY: Wiley, 1978, pp. 135–285.

[3] Driggers, R., P. Cox, and T. Edwards, *Introduction to Infrared and Electro-Optical Systems,* Boston, MA: Artech House, 1999, pp. 39–57.

[4] Deltoro, V. *Electrical Engineering Fundamentals,* 2nd Ed., Englewood Cliffs, NJ: Prentice-Hall, 1986, p. 177.

[5] Oppenhiem, A., A. Willsky, and I. Young, *Signals and Systems,* Englewood Cliffs, NJ: Prentice-Hall, 1978, pp. 161–226.

[6] Gaskill, J. *Linear Systems, Fourier Transforms, and Optics,* New York, NY: Wiley, 1978, pp. 305–307.

[7] Hecht, E., and A. Zajak, *Optics,* Reading, MA: Addison-Wesley, 1979, p. 175.

[8] Dereniak, E., and G. Boreman, *Infrared Detectors and Systems,* New York, NY: Wiley, 1996, p. 513.

[9] Holst, G., *Electro-Optical Imaging System Performance,* Winter Park, FL: JCD Publishing, 1995, p. 82.

[10] Waldman, G., and J. Wootton, *EO System Performance Modeling,* Norwood, MA: Artech House, 1993, p. 170.

[11] Goodman, C., "The Relative Merits of LEDs and LCDs," *Proceedings for Information Display,* Vol. 16, No. 1, 1975.

[12] Overington, I., *Vision and Acquisition,* New York, NY: Crane Russak, and Co., 1976.

[13] Vollmerhausen, R., and R. Driggers, *Analysis of Sampled Imaging Systems,* Bellingham, WA: SPIE Press, 2000.

[14] Vollmerhausen, R., "Display of Sampled Imagery," *IRIA-IRIS Proceedings: Meeting of the IRIS Specialty Group on Passive Sensors,* Vol. 1, 1990, pp. 175–192.

[15] Vollmerhausen, R., "Impact of Display Modulation Transfer Function on the Quality of Sampled Imagery," *SPIE Aerospace/Defense Sensing and Controls,* Vol. 2743, 1996, pp. 12–22.

[16] Vollmerhausen, R., R. Driggers, and B. O'Kane, "The Influence of Sampling on Target Recognition and Identification," *Optical Engineering,* Vol. 38, No. 5, May 1999.

[17] Vollmerhausen, R., et al., "Staring Imager Minimum Resolvable Temperature (MRT) Measurements Beyond The Sensor Half Sample Rate," *Optical Engineering,* Vol. 37, No. 6, January 1998, p. 1763.

[18] Vollmerhausen, R., and R. Driggers, "NVTHERM: Next Generation Night Vision Thermal Model," *1999 IRIS Passive Sensors Proceedings,* 440000-121-X(1), Vol. 1, IRIA Center, ERIM, Ann Arbor, MI, pp. 121–134.

[19] Vollmerhausen, R., and R. Driggers, "Modeling the Target Acquisition Performance of Staring Array Sensors," *1998 IRIS Passive Sensors Proceedings,* 440000-82-X(1), Vol. 1, IRIA Center, ERIM, Ann Arbor, MI, pp. 211–224.

[20]   Vollmerhausen, R., R. Driggers, and B. O'Kane, "The Influence of Sampling on Recognition and Identification Performance," *Optical Engineering*, Vol. 36, No. 5, May 1999.

[21]   Edde, B., *Radar Principles, Technology, Applications*, Englewood Cliffs NJ: Prentice-Hall, 1993, p. 628.

[22]   Levanon, N., *Radar Principles*, New York, NY: Wiley, 1988, p. 106.

[23]   Oliver, C., and S. Queqan, *Understanding Synthetic Aperture Radar Images*, Norwood, MA: Artech House, 1998.

5

Information Extraction Measures

In this chapter, we begin the development of surveillance and reconnaissance (S&R) models and performance predictors. The goal of the S&R process is to provide information. We thus begin by discussing measures of information extraction, both direct and indirect. We describe these measures as information extraction performance measures or metrics. Information extraction metrics are the measures of information extraction that S&R models attempt to predict. They include direct performance measures and performance prediction measures. This chapter defines and provides examples for most of the commonly used information extraction metrics. Particular attention is paid to the statistical properties of the variables as a means to better understand their behavior in model development. To this end, a brief review of the theory of signal detection (TSD) is included.

Direct performance measures are typically defined as probabilities. For example, "on an image of quality X, the probability of correctly identifying an M-60 tank is Y." The probability Y can be considered a likelihood. Given n images of quality x and m observers, the tank will be correctly identified Y proportion of times. As an equation:

$$Y = \frac{C}{(n \times m)} \quad \text{where} \tag{5.1}$$

Y is the probability of correct identification, C is the number of correct identifications, n is the number of images, and m is the number of observers.

A number of factors other than image quality or the imaging process affect information extraction performance. The target and its environment affect performance. Individual observer differences such as visual acuity and training play a role. Within-observer effects can be even more important. These effects are modeled within the context of signal detection theory.

Although probability of task performance is the most common direct measure of information extraction, other measures also have useful applications. Chief among these are time and range metrics. How long, on average, does it take to recognize a tank on an image of quality X? At what range can a tank be detected with system Y?

Information in the sense of classic information theory is an objective measure derived from physical quality metrics or system design parameters. Information density has been used as a predictor of National Imagery Interpretability Rating Scale (NIIRS) values [1]. Other physically based metrics include difference measures such as just-noticeable-difference estimates based on models of human visual performance [2].

Predictive measures include the previously mentioned NIIRS, observer estimates of task satisfaction, and relative quality scales. In the sections that follow, each of these information extraction variables is defined in further detail and discussed in the context of statistical behavior. Statistical behavior includes response variability, reliability, and linearity.

5.1 Direct Performance Measures

The goal of S&R modeling in the context of this book is to predict overall imaging system and observer information extraction performance. In the process of developing a model, it is desirable to validate that model. Validation is the process of objectively demonstrating the accuracy of the model [3]. Direct performance measurement provides the basis for direct validation. As we will describe, however, direct performance measures are difficult to use in practice.

5.1.1 Probability of Task Performance

Information extraction tasks have generally been defined along a hierarchy ranging from detection to identification. Table 5.1 provides some examples. Detection implies some level of identification; otherwise, no response would be made to the detected object. To detect a vehicle requires some recognition that the object in question is at least potentially a vehicle. Beyond the first stage of detection, there can be several levels of classification and iden-

Table 5.1
Recognition Hierarchy

Discrimination Level	M-60A2 Tank	Object F/A-18	Spruance Class DD*
Detect	Vehicle	Aircraft	Ship
Classify	Tank	Fighter	Combatant
Classify	Medium tank	Medium fighter	Medium combatant
Identify	M-60	F-18	DD
Identify	M-60A2	F/A-18	Spruance Class DD

*Destroyer.

tification. Recognition and classification imply the same general level of naming. "Recognizing" as a tank implies "classification" as a tank.

In the example given, additional levels of classification and identification might have been inserted. For example, prior to classifying the M-60 as a tank, it might have been classified as an armored fighting vehicle (AFV). As one moves from detection to identification, more information is implied. Assuming experience or other sources of intelligence, identification as an M-60A2 provides more information on capabilities than identification simply as a tank. The additional information comes with the cost of a need for additional image quality.

The classification hierarchy is not restricted to military objects; it applies to any set of natural or cultural features. For example, tree, coniferous tree, pine tree, yellow pine tree. Any recognition guide or key provides numerous examples.

It is apparent that as one moves down the hierarchy, the level of quality required to accomplish the task increases. Identification is more difficult and requires higher quality than detection. What is perhaps less apparent is that at each level in the hierarchy, there are a number of possible alternatives. For example, consider the hierarchy shown in Figure 5.1.

Alternative choices are available at each level of the hierarchy. Alternatives at each level are not at the same level of difficulty; however, they depend on the actual object present. Some armored personnel carriers (APCs) and self-propelled guns (SPGs) look very similar to a tank, others do not. The viewing aspect angle also affects the level of difficulty. An APC that looks like a tank in the plan view (from above) may be distinctively different from a side view. For vehicles, frontal views tend to be more difficult than side or plan views.

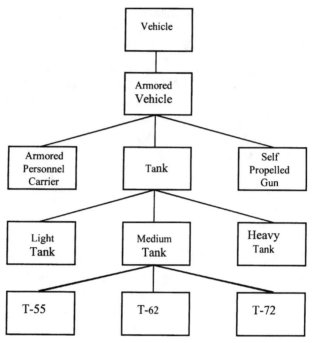

Figure 5.1 Recognition hierarchy.

Information extraction performance is what we want to predict with a model, but, for several reasons, it is not easy to measure. It is generally accepted that an unbiased measurement of performance can only be obtained once from a given image of a scene. Once an observer has viewed a scene, learning has taken place. Performance on successive scenes is thus based in part on the quality of the successive scenes and in part on learning from the first scene. This means that images of several different scenes are needed to accurately measure performance. As interest proceeds to higher levels of identification, object alternatives need to be represented. From Figure 5.1, images are needed not only of T-62 tanks, but of other tank models, APC, and SPG models as well. Depending on the sensor of interest, image scenes may need to be imaged from several different aspect angles. The image sample needs to show a range of imaging conditions such as sun angle for visibility, weather and time of night for IR, and target background for SAR. Finally, the operational scenario may need to be reproduced with fidelity. To the extent that fatigue, boredom, rewards and penalties for correct and incorrect responses, and environmental factors affect performance, they need to be replicated in the test environment if performance is to be truly measured.

5.1.2 Completeness and Accuracy

Probabilities of detection, classification, and identification are positive measures in that they represent correct responses. Probabilities of incorrect responses are also used as direct performance measures. A false detection is termed a false alarm. A nontarget is incorrectly called a target. Incorrect classification and identification responses are classed as misidentifications. False alarm and misidentification probabilities are computed in the same manner as probabilities of detection and identification.

Probability measures represent performance on a set of n observations. A detection probability (P_D) indicates the probability that any one of a number of like objects will be detected. In practice, detection probabilities are typically measured by showing observers a sample of objects, some of which are targets and some not. A related measure that has been used frequently is completeness [4]. Completeness is the percentage of objects present (in an image or set of images) that are correctly detected in an evaluation or performance trial. It is thus defined as

$$C(\%) = 100 \times \text{Correct detections/total number of targets} \qquad (5.2)$$

Completeness is affected by both image quality and the extent to which the observer scans and attends to the content of the total image. An object may be missed because it is not detectable or because it is simply not noticed. The process of search is discussed in more detail in Chapter 9.

When completeness is used to assess detection performance, accuracy is the measure used to assess identification performance. Accuracy is defined as

$$A(\%) = 100 \times \text{Correct Identifications/Total Identification responses} \qquad (5.3)$$

Whereas the probability of classification or identification is based on the proportion of objects present correctly identified, accuracy is based on the proportion of responses correctly made. A conservative observer may have low completeness (few detections) and high accuracy (most detections correctly identified) because of responding only to obvious targets. That same observer in the same situation would have a low probability of both detection and identification because both probability of detection and identification is computed against the total target population, not just those to which a response was made.

5.1.3 Statistical Properties of Direct Measures

Over a full image quality continuum, the relationship between probability measures and quality is nonlinear. Figure 5.2 shows the typical function. The relationship is nonlinear because there is some range of quality where performance is 0 and another where it is perfect (1.00). As these ranges are approached, performance is asymptotic. At some interval between the two ranges, performance is linear with quality.

It is also the case that response variability is highly correlated with mean performance. Figure 5.3 shows a set of target identification data [5]. Observers attempted to identify IR images of military vehicles under various levels of blur. Because the scoring of responses is binary (correct or incorrect), the standard deviation is a function of mean performance. Variability is at a maximum when half the responses are correct and at a minimum when they are all correct or all incorrect. From Figure 5.3, an average variability of ~0.35 can be estimated assuming a rectangular distribution of means from 0 to 100. This is slightly more than one third of the total response scale range.

Reliability is another measure that can be used to characterize performance prediction variables. Reliability is a correlation measure denoting the expected correlation between two replications of the same study. A high reliability value indicates that the results of a study (results of performance

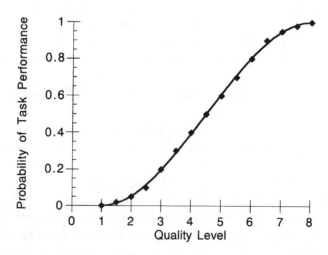

Figure 5.2 Typical quality/utility relationship.

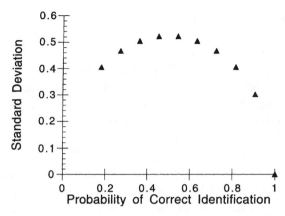

Figure 5.3 Relationship between mean and standard deviation for identification data.

measurement) are stable and that conclusions drawn from results are stable. Reliability can be measured directly or it can be estimated from:

$$r = (MS_{between\ ratings} - MS_{within\ ratings}) / MS_{between\ ratings} \qquad (5.4)$$

where MS is the mean square value computed from analysis of variance.

Both the performance metric and the characteristics of the measured data affect reliability. If all of the observers are in agreement (make similar responses or show low within-rating variability) and all the data show a good range of performance (high between-rating), reliability is high. If between-rating variability is low (all the data show similar performance), reliability is low. For the data shown in Figure 5.3, the reliability value was 0.84.

From a theoretical point of view, the nonlinear relationship between performance and quality can be expressed by a logit function [6]. The logit function derives from log-linear modeling. Log-linear modeling was developed to analyze contingency table data such as shown in Table 5.2. In the

Table 5.2
Contingency Table

Voting turnout	Membership	
	Belong to one or more civic groups	Belong to none
Voted	673	234
Not voted	126	879

context of S&R modeling, a response is made or it is not made. The response is correct, or it is incorrect. Log-linear modeling is similar to regression except that, rather than predicting a value on an ordinal or ratio scale, the model predicts odds of a response. Odds in this context is the same as in betting ("The odds on Lucky Larry in the seventh are 2:1"). The odds represent the ratio of two frequencies. In the case of Lucky Larry, they are the expected frequencies of not winning (2) and winning (1). The odds on Betsy in the same race is 8:1. The loss/win ratio for Betsy is 8:1. The odds of rolling a seven (with two dice) is 1:6.

In general log-linear modeling, no distinction is made between dependent and independent variables. The data is treated in terms of a contingency table, and the goal is to predict the cell frequencies. With the logit model, one variable is chosen as the dependent variable. In a detection study, the frequency of correct detections would be the dependent variable. The logit model predicts the odds of a correct detection.

In standard linear regression

$$E(Y \mid x_1) = a + bx_1 \tag{5.5}$$

where $E(Y \mid x_1)$ is the conditional mean of Y given x_1, a is the regression constant or intercept, and b is the regression coefficient. Y is the dependent variable and can assume any value on the Y scale depending on the values of x.

In logistic regression, the dependent variable assumes one of two values, 1 (correct) or 0 (incorrect). With logistic regression

$$E(Y \mid x_1) = p_{X_1} \tag{5.6}$$

where p_{X_1} is the probability that $Y = 1$ given that $X_1 = x_1$.

This linear form shown in (5.5) suffers from the fact that, although the right side of the equation can take any value, the left side must be a value between 0 and 1. In logistic regression, a nonlinear transform of the dependent variable is applied. This transform is

$$\log\left(\frac{p_{x_1}}{1 - p_{x_1}}\right) \tag{5.7}$$

which is the log of the odds that $Y = 1$. Odds is the probability of $Y = 1$ divided by 1 minus that probability given x_1. The original linear regression can be rewritten as

$$\log\left(\frac{p_{x_1}}{1 - p_{x_1}}\right) = b_0 + bx_1 \qquad (5.8)$$

The model assumes that the log odds of the dependent variable change linearly with changes in X_1. This is shown with a hypothetical set of data in Figure 5.4. The left side of the figure shows the fit to the original probabil-

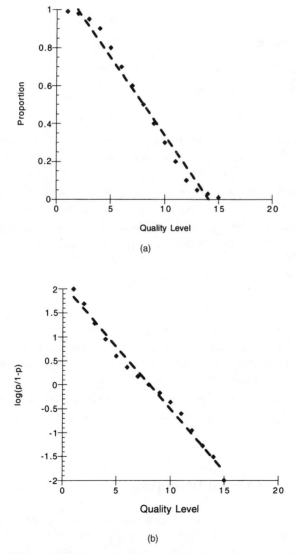

Figure 5.4 Effect of logit function: (a) proportion, and (b) logit.

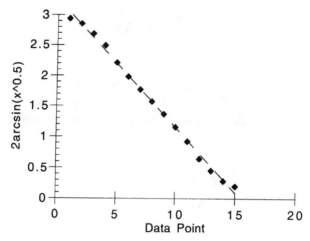

Figure 5.5 Effect of arcsin transform.

ity data. The right side shows the fit after the odds or logit transform. The fit is not perfect, which is often the case in real life. The logit model works best where there is a complete continuum of probabilities. Where there is not, other regression models may work as well or better. For example, in Figure 5.4(a), a linear fit for points 4 through 12 does as well. A nonlinear transform may also be applied to the data. The arcsin transform has much the same effect as the logit transform [7]. It is of the form

$$x'_{ijk} = 2 \arcsin \sqrt{x_{ijk}} \qquad (5.9)$$

where x'_{ijk} is the transform of the original proportion x_{ijk}.

Figure 5.5 shows an arcsin transform applied to the data in Figure 5.4(a). It appears to perform slightly better than the logit transform in terms of achieving linearity. More important is the effect of the transform in achieving homogeneity of error variance. For certain statistical tests, variance should be independent of the mean (i.e., constant regardless of the value of the mean).

5.2 Theory of Signal Detection

Signal detection theory is a body of analysis developed over roughly a 20-year period beginning in 1945. It combines decision theory and the statistical properties of signals. Several comprehensive texts have been written on the subject [8–10]. The intent here is not to present the details of TSD but rather to show how it applies to the subject of this chapter.

In a simple forced choice detection study, four outcomes are possible at each trial. Figure 5.6 shows the possibilities. They are labeled A through D. The same approach can be extended to a forced recognition or identification study.

TSD postulates the problem in terms of two distributions of events, noise and signal (or target) plus noise. This is diagrammed in Figure 5.7 in terms of event distributions. Some events are noise (nontargets), some are signal plus noise. Where the two distributions do not overlap, the perfect observer would make no mistakes and would always detect a target and correctly reject a nontarget. Conceptually, there are some targets that are obvious and some that are difficult to separate from noise (or target-like objects). Targets thus exist along a continuum, with most in the middle of the distribution and some at both ends of the continuum. Similarly, some nontargets may closely resemble targets and others obviously do not. Again, a continuum of "target likeness" exists. In the area where the distributions overlap, errors can and do occur. True targets may be rejected and false targets may be accepted.

The greater the similarity between noise and signal plus noise, the greater the overlap of the two distributions (Figure 5.7). In TSD, the separation between the two distributions is defined as d', the difference between two means divided by the common standard deviation. Increasing d' makes the detection task easier.

The observer can vary the point along the x axis where a signal, as opposed to a noise, response is made. This is the observer's decision threshold

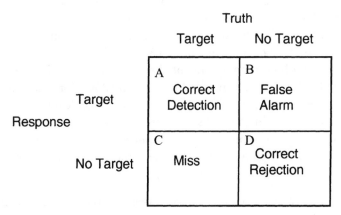

Figure 5.6 Decision alternatives.

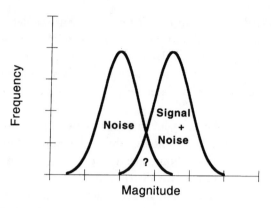

Figure 5.7 Distribution of noise and signal plus noise.

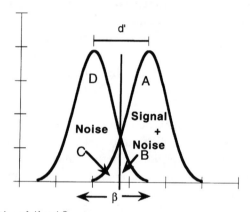

Figure 5.8 Illustration of d' and β.

criterion or β (Figure 5.8). A very conservative observer will move the decision threshold to the right. This observer will incur low false alarm rates at the expense of few detections. Similarly, by moving the criterion to the left, high detection rates will occur at the expense of high false alarm rates. Figure 5.8 also shows the four possible outcomes (A through D) from Figure 5.6.

In TSD, data is commonly presented in terms of a receiver operating characteristic (ROC) curve. Figure 5.9 shows two ROCs for Gaussian distributions with equal variance.[1] The ROC shows all possible outcomes for a given noise and signal plus noise distribution as the decision criterion, β, is changed. The value d' is represented by distance along the negative diagonal (Figure 5.9). The negative diagonal is the perpendicular to the line con-

---

1. There is no requirement that the distributions be Gaussian. They may be exponential, binomial, or any other of a variety of distributions.

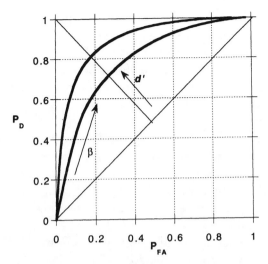

Figure 5.9 Receiver operating characteristic (ROC) curves.

necting values $P_D/P_{FA} = 0$ and $P_D/P_{FA} = 1$ and the value β is a point on the ROC. The conservative β falls toward the lower left of the ROC, the β selected to maximize detections falls to the upper right.

The discussion of TSD to this point has covered only the two-event/two-choice situation. The ROC can be extended to a situation where multiple responses to the same stimulus set are made. Multiple targets are present, some of which are detected. Some number of false alarms will occur over the stimulus area (the image). By varying β, what is called a *free* response operating characteristic (FROC) curve can be generated. This curves (Figure 5.10)

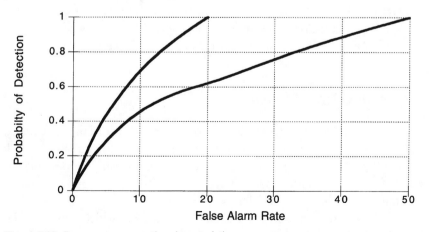

Figure 5.10 Free response operating characteristic curves.

plot probability of detection (of all of the targets present) against false alarm rate (false alarms per unit area) as a function of β [11,12]. In other respects, FROCs can be considered as an extension of TSD to the multiple opportunity case. FROCs are particularly useful in studies of search performance where targets (and false alarms) are distributed across large image areas.

In the context of this book, d′ is roughly a measure of image quality or interpretability. As d′ increases, higher detection rates can be achieved with relatively low false alarm rates. For a given d′, however, there are many possible values of β. Several factors can affect the observer's decision criterion. Observers attempt to maximize success in their own mind. The subjective cost of a missed detection or a false alarm varies β. The expected or observed frequency of targets as opposed to nontargets also varies β, as do a variety of other psychological factors [8].

In measuring detection and recognition performance, the S&R literature has, with a few exceptions, tended to ignore TSD. At best, the typical study may strive to achieve a "neutral" β (i.e., a β on the negative diagonal of the ROC, assuming Gaussian distributions). The neutrality hypothesis is seldom tested. As a result, it is nearly impossible to compare data across studies because both d′ and β may differ. A performance prediction for a system of given d′ performance can vary substantially in terms of correct responses unless β is somehow controlled.

Perhaps one of the reasons why TSD has not been applied more extensively is the amount of data that must be collected to fully define the ROC. For a target present/not present type of study (yes/no responses), β must be systematically varied over the range of the ROC. If confidence data is collected along with the yes/no responses, it can be used to estimate the ROC by plotting results as a function of confidence, but typically with greater uncertainty. Forced choice methods can be used, and with proper instructions and procedures, the assumption of a neutral β can be made. There is some question as to how successful this method is, however [13].

5.3 Range/Time Measures

For some purposes, it is desirable to determine the time or range at which a desired level of performance (detection or identification) takes place. A tank gunner needs to identify an enemy tank in sufficient time to aim and fire before being fired upon. A weapons officer would like to acquire the target at a range exceeding that of defensive weapons.

For such applications, one measures the range at which a target is detected or identified, or the time from initial presentation required to detect or identify the target. Both time and range typically assume non-normal distributions. There is some minimum time below which even the fastest observer cannot perform. It is also the case that a few observers or observing situations require an inordinate amount of time before a response is made. Figure 5.11 shows an example. The data was taken from a study where observers were asked to identify pre-located targets on aerial photography [14]. Figure 5.11(a) shows response times for bomber aircraft at a particular level of contrast and resolution. The distribution is decidedly skewed. Where homogeneity of variance is desired for purposes of analysis, a transform of the data is required. Figure 5.11(b) shows the data after a log transform has been applied to the response times. The data is less skewed.

In the case of range, there are two factors at work. The first is again the time required to respond, the second is the increase in target subtense as range decreases. A log transform generally stabilizes variance for range, just as it does with time.

Time has also been used as a performance metric in conjunction with detection or recognition metrics. Bennett, et al. [15], for example, measured completeness as a function of time. Birnbaum [16] measured completeness and accuracy as a function of time. In both cases, the findings of TSD were shown, although not in the context of TSD. A sample of data from the Bennett study is shown in Figure 5.12. Performance improved over time with diminishing returns. Airfield performance reached 100% completeness at 50 seconds, but 80% completeness was achieved in half that time. Performance on industrial targets never rose above 60% even at 200 seconds and was almost at that level at 100 seconds. In both cases, a polynomial fit to the data showed correlations in excess of 0.90. The Birnbaum data showed accuracy decreasing over time. Both sets of results are consistent with TSD in that over time, and with a finite target population, the observers' decision criterion becomes more liberal in order to continue responding.

5.4 Performance Estimate Measures

Performance estimate measures represent observer estimates of information extraction performance. NIIRS ratings are an indication of the level (difficulty) of information that can be extracted from a given image [17]. Increasing levels indicate greater amounts or specificity of information with an implied requirement for increased quality. Task satisfaction confidence (TSC) ratings are the observers' estimates of their ability to perform a specific information extraction task on an image. Finally, a variety of psychophysical scaling techniques are used to define relative quality or interpretability.

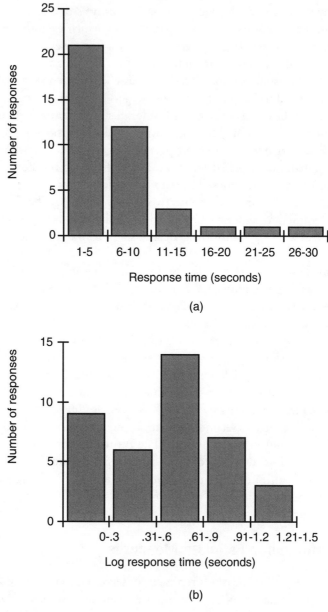

Figure 5.11 Distribution of response time data (a) with and (b) without log transform. (Data from [14].)

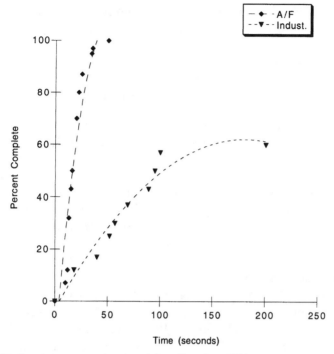

Figure 5.12 Completeness as a function of time. (Data from [15].)

In the previous sections, some of the difficulties inherent in directly measuring performance were discussed. Performance estimate techniques tend to be less resource-intensive. Different versions of the same scene can, in theory, be rated with no biasing effects. Task learning is typically much less of an effect. There is no need to attempt to physically replicate the operational environment, although there may be a need to describe the environment in order to guide the observers in making their estimates. As will be shown, however, not all aspects of performance can be estimated. Despite the variability in measurement using direct performance measures, there is sometimes a decision-maker bias toward the use of such measures. Performance estimates have the appearance of greater subjectivity. Despite this perception, there is no evidence to support the relative validity of the two types of measures. It has, to the authors' knowledge, proved impossible to accurately measure operational performance. Validation studies comparing measured performance to estimated performance (in laboratory settings) have shown good agreement in the sense of high correlations. As will be shown, IAs collectively are good predictors of at least some measures of performance.

Thus, one is forced to rely on face validity[2] and accept that the strong correlation between direct and predictive measures validates the performance estimating measures.

5.4.1 National Imagery Interpretability Rating Scale

NIIRS was developed in response to the inability of simple measures of physical image quality to successfully communicate image interpretability. It was apparent that the relationship between physical image quality and information extraction performance was multidimensional. With multiple system designs, image scale was no longer a good predictor of performance. Resolution, as measured by tri-bar targets on the ground, was also not a good predictor because other physical measures also affected performance. Simple relative scales (e.g., poor, fair, good, excellent) failed to account for major system differences. What was a good image for one imaging system might be a poor image for another.

Originally developed in the early 1970s by a government and contractor team operating under the auspices of the U.S. government's Imagery Resolution and Reporting Standards (IRARS) Committee,[3] NIIRS has been used since then by the intelligence community to communicate the interpretability of imagery [17]. The goal in the original development of NIIRS was to develop a scale that would communicate what information could and could not be extracted from a specific image. In the initial development process, a large sample of interpretation tasks was rated in terms of relative difficulty by a group of IAs. The tasks ranged in difficulty from detection of large facilities to identification of small target details. All of the tasks were related to intelligence applications and dealt with objects and features of military interest. A sample of visible imagery of varying and known quality was provided to another group of IAs, and they were asked to determine, on each image, the most difficult task that could "just be accomplished."

The final scale was defined in terms of variations in resolution. Each level of the scale represented a doubling or halving of resolution ($\log_2$ resolution). A set of tasks or criteria was selected at each of 10 levels (based on the IAs' judgments as to what could be accomplished) to define a 0–9 scale. The tasks at each level were defined for five military orders-of-battle (air,

---

2. Validity is defined as the correlation between predicted and observed performance. As an example, a test used to select auto salespersons should correlate with the number of autos sold by the same persons. Face validity simply means that the test appears to correlate with actual performance.
3. IRARS was originally established to standardize the rating of tri-bar resolution targets.

ground, naval, missile, electronics). The tasks or criteria defined specific objects and levels of interpretation (detect, recognize, identify).

After the original NIIRS was introduced, it became apparent that IAs had difficulty providing ratings when the objects referenced in the criteria were not present in the scene. In addition, over time, many of the referenced objects fell out of use and were thus not familiar to IAs. Finally, as the use of radar and IR systems increased, the need for a scale to deal with those systems became evident. Different characteristics and contributions of radar and IR made it evident that one could not simply apply the visible scale. For these reasons, the original visible scale was revised and scales developed for IR and radar. A cultural criteria category was added to the scales for cases where military equipment was not present. Tables 5.3 through 5.5 show the current versions of each scale. The most recent version of the visible scale was developed in 1994. The current version of the IR scale was published in 1996 and the current radar scale in 1999. The scales are currently maintained by the National Imagery and Mapping Agency.

The introduction and growing utility of commercial satellite systems led to the development of two additional NIIRS. The first was the Civil NIIRS [18], which was developed in response to requirements of civilian agencies, such as the U.S. Geological Survey, the U.S. Forest Service, and the Bureau of Land Management [19]. Users in the civil sector could not use the military criteria; they needed natural and additional cultural criteria. The Civil NIIRS criteria categories—natural, agricultural, and urban/industrial—were established to meet these requirements. The scale is shown in Table 5.6 and applies only to visible imagery. The Civil NIIRS is calibrated to the current visible NIIRS, meaning that an image that will support a visible NIIRS 5 task will also support a Civil NIIRS 5 task.

Increasing development and use of multispectral systems led to a multispectral version of the NIIRS [20]. The initial attempt was termed the MS IIRS and is shown in Table 5.7. The MS IIRS covered cultural and natural criteria and had only seven levels as opposed to the 10 for the other NIIRS. A 10-level MS NIIRS is currently under development.

Currently, NIIRS are developed in a three-step process [21]. First, a sample of imagery with as much quality diversity as possible is assembled for each of the scale criteria categories (e.g., air, ground, cultural). IAs then scale the imagery on a unidimensional interpretability continuum (subjective quality scale, SQS). Interval scaling between marker images of 0 and 100 is the most common technique, but magnitude estimation[4] has also been employed.

---

4. Magnitude estimation is a scaling technique where images are scaled using a ratio scale relative to a single standard image. The image to be rated may be better or worse than the standard. This technique simplifies the process of rating in softcopy.

Table 5.3
Visible National Imagery Interpretability Rating Scale—March 1994

Rating Level 0

Interpretability of the image is precluded by obscuration, degradation, or very poor resolution.

Rating Level 1

Detect a medium-sized port facility and/or distinguish between taxiways and runways at a large airfield.

Rating Level 2

Detect large hangars at airfields.

Detect large static radars (e.g., AN/FPS-85, COBRA DANE, PECHORA, HENHOUSE).

Detect military training areas.

Identify an SA-5 site based on road pattern and overall site configuration.

Detect large buildings at a naval facility (e.g., warehouses, construction hall).

Detect large buildings (e.g., hospitals, factories).

Rating Level 3

Identify the wing configuration (e.g., straight, swept, delta) of all large aircraft (e.g., 707, CONCORD, BEAR, BLACKJACK).

Identify radar and guidance areas at a SAM site by the configuration, mounds, and presence of concrete aprons.

Detect a helipad by the configuration and markings.

Detect the presence/absence of support vehicles at a mobile missile base.

Identify a large surface ship in port by type (e.g., cruiser, auxiliary ship, noncombatant/merchant).

Detect trains or strings of standard rolling stock on railroad tracks (not individual cars).

Rating Level 4

Identify all large fighters by type (e.g., FENCER, FOXBAT, F-15, F-14).

Detect the presence of large individual radar antennas (e.g., TALL KING).

Identify, by general type, tracked vehicles, field artillery, large river crossing equipment, wheeled vehicles when in groups.

Detect an open missile silo door.

Determine the shape of the bow (pointed or blunt/rounded) on a medium-sized submarine (e.g., ROMEO, HAN, Type 209, CHARLIE II, ECHO II, VICTOR II/III).

Identify individual tracks, rail pairs, control towers, switching points in rail yards.

Rating Level 5

Distinguish between a MIDAS and a CANDID by the presence of refueling equipment (e.g., pedestal and wing pod).

Identify radar as vehicle-mounted or trailer-mounted.

Identify, by type, deployed tactical SSM systems (e.g., FROG, SS-21, SCUD).

Table 5.3 (cont.)

Distinguish between SS-25 mobile missile TEL and missile support vans (MSVs) in a known support base, when not covered by camouflage.

Identify TOP STEER or TOP SAIL air surveillance radar on KIROV-, SOVREMENNY-, KIEV-, SLAVA-, MOSKVA-, KARA-, or KRESTA-II-class vessels.

Identify individual rail cars by type (e.g., gondola, flat, box) and/or locomotives by type (e.g., steam, diesel).

Rating Level 6

Distinguish between models of small/medium helicopters (e.g., HELIX A from HELIX B from HELIX C, HIND D from HIND E, HAZE A from HAZE B from HAZE C).

Identify the shape of antennas on EW/GCI/ACQ radars as parabolic, parabolic with clipped corners, or rectangular.

Identify the spare tire on a medium-sized truck.

Distinguish between SA-6, SA-11, and SA-17 missile airframes.

Identify individual launcher covers (8) of vertically launched SA-N-6 on SLAV-class vessels.

Identify automobiles as sedans or station wagons.

Rating Level 7

Identify fitments and fairings on a fighter-sized aircraft (e.g., FULCRUM, FOXHOUND).

Identify ports, ladders, vents on electronic vans.

Detect the mount for antitank guided missiles (e.g., SAGGER on BMP-1).

Detect details of the silo door hinging mechanism on Type III-F, III-G, II-H launch silos and type III-X launch control silos.

Identify the individual tubes of the RBU on KIROV-, KARA-, KRIVAK-class vessels.

Identify individual rail ties.

Rating Level 8

Identify the rivet lines on bomber aircraft.

Detect horn-shaped and W-shaped antennas mounted atop BACK TRAP and BACKNET radars.

Identify a hand-held SAM (e.g., SA-7/14, REDEYE, STINGER).

Identify joints and welds on a TEL or TELAR.

Detect winch cables on deck-mounted cranes.

Identify windshield-wipers on a vehicle.

Rating Level 9

Differentiate cross-slot from single-slot heads on aircraft skin panel fasteners.

Identify small, light-toned ceramic insulators that connect wires of an antenna canopy.

Identify vehicle registration numbers (VRN) on trucks.

Identify screws and bolts on missile components.

Identify braid of ropes (3–5 inches in diameter).

Detect individual spikes in railroad ties.

Table 5.4
Infrared National Imagery Interpretability Rating Scale—April 1996

Rating Level 0

Interpretability of the imagery is precluded by obscuration, degradation, or very poor resolution.

Rating Level 1

Distinguish between runways and taxiways on the basis of size, configuration or pattern at a large airfield.

Detect a large (e.g., greater than 1 km^2) cleared area in dense forest.

Detect large ocean-going vessels (e.g., aircraft carrier, super-tanker, KIROV) in open water.

Detect large areas (e.g., greater than 1 km^2) of marsh/swamp.

Rating Level 2

Detect large aircraft (e.g., C-141, 707, BEAR, CANDID, CLASSIC).

Detect individual large buildings (e.g., hospitals, factories) in an urban area.

Distinguish between densely wooded, sparsely wooded and open fields.

Identify an SS-25 base by the pattern of buildings and roads.

Distinguish between naval and commercial port facilities based on type and configuration of large functional areas.

Rating Level 3

Distinguish between large (e.g., C-141, 707, BEAR, A-300 AIRBUS) and small aircraft (e.g., A-4, FISHBED, L-39).

Identify individual thermally active flues running between the boiler hall and smoke stacks at a thermal power plant.

Detect a large air warning radar site based on the presence of mounds, revetments, and security fencing.

Detect a driver training track at a ground forces garrison.

Identify individual functional areas (e.g., launch sites, electronics area, support area, missile handling area) of an SA-5 launch complex.

Distinguish between large (e.g., greater than 200m) freighters and tankers.

Rating Level 4

Identify the wing configuration of small fighter aircraft (e.g., FROGFOOT, F-16, FISHBED).

Detect a small (e.g., 50m^2) electrical transformer yard in an urban area.

Detect large (e.g., greater than 10m diameter) environmental domes at an electronics facility.

Detect individual thermally active vehicles in garrison.

Detect thermally active SS-25 MSVs in garrison.

Identify individual closed cargo hold hatches on large merchant ships.

Rating Level 5

Distinguish between single-tail (e.g., FLOGGER, F-16, TORNADO) and twin-tailed (e.g., F-15, FLANKER, FOXBAT) fighters.

Identify outdoor tennis courts.

Table 5.4 (cont.)

Identify the metal lattice structure of large (e.g., approximately 75m) radio relay towers.

Detect armored vehicles in a revetment.

Detect a deployed transportable electronics tower (TET) at an SA-10 site.

Identify the stack shape (e.g., square, round, oval) on large (e.g., greater than 200m) merchant ships.

Rating Level 6

Detect wing-mounted stores (i.e., ASM, bombs) protruding from the wings of large bombers (e.g., B-52, BEAR, BADGER).

Identify individual thermally active engine vents atop diesel locomotives.

Distinguish between a FIX FOUR and FIX SIX site based on antenna pattern and spacing.

Distinguish between thermally active tanks and APCs.

Distinguish between a 2-rail and 4-rail SA-3 launcher.

Identify missile tube hatches on submarines.

Rating Level 7

Distinguish between ground attack and interceptor versions of the MIG-23 FLOGGER based on the shape of the nose.

Identify automobiles as sedans or station wagons.

Identify antenna dishes (less than 3m in diameter) on a radio relay tower.

Identify the missile transfer crane on an SA-6 transloader.

Distinguish between an SA-2/CSA-1 and a SCUD-B missile transporter when missiles are not loaded.

Detect mooring cleats or bollards on piers.

Rating Level 8

Identify the RAM airscoop on the dorsal spine of FISHBED J/K/L.

Identify limbs (e.g., arms, legs) on an individual.

Identify individual horizontal and vertical ribs on a radar antenna.

Detect closed hatches on a tank turret.

Distinguish between fuel and oxidizer multisystem propellant transporters based on twin or single fitments on the front of the semitrailer.

Identify individual posts and rails on deck edge life rails.

Rating Level 9

Identify access panels on fighter aircraft.

Identify cargo (e.g., shovels, rakes, ladders) in an open-bed, light-duty truck.

Distinguish between BIRDS EYE and BELL LACE antennas based on the presence or absence of small dipole elements.

Identify turret hatch hinges on armored vehicles.

Identify individual command guidance strip antennas on an SA-2/CSA-1 missile.

Identify individual rungs on bulkhead mounted ladder.

Table 5.5
Radar National Imagery Interpretability Rating Scale—August 1992

Rating Level 0

Interpretability of the imagery is precluded by obscuration, degradation, or very poor resolution.

Rating Level 1

Detect the presence of aircraft dispersal parking areas.

Detect a large cleared swath in a densely wooded area.

Detect, based on presence of piers and warehouses, a port facility.

Detect lines of transportation (either road or rail, but do not distinguish between).

Rating Level 2

Detect the presence of large (e.g., BLACKJACK, CAMBER, COCK, 707, 747) bombers or transports.

Identify large phased array radars (e.g., HENHOUSE, DOGHOUSE) by type.

Detect a military installation by building pattern and site configuration.

Detect road pattern, fence, and hardstand configuration at SSM launch sites (missile silos, launch control silos) within a known ICBM complex.

Detect large noncombatant ships (e.g., freighters or tankers) at a known port facility.

Identify athletic stadiums.

Rating Level 3

Detect medium-sized aircraft (e.g., FENCER, FLANKER, CURL, COKE, F-15).

Identify an ORBITA site on the basis of a 12-m dish antenna normally mounted on a circular building.

Detect vehicle revetments at a ground forces facility.

Detect vehicles/pieces of equipment at a SAM, SSM, or ABM fixed missile site.

Determine the location of the superstructure (e.g., fore, amidships, aft) on a medium-sized freighter.

Identify a medium-sized (approximately six-track) railroad classification yard.

Rating Level 4

Distinguish between large rotary-wing and medium fixed-wing aircraft (e.g., HALO versus CRUSTY transport).

Detect recent cable scars between facilities or command posts.

Detect individual vehicles in a row at a known motor pool.

Distinguish between open and closed sliding roof areas on a single bay garage at a mobile base station.

Identify square bow shape of ROPUCHA class (LST).

Detect all rail/road bridges.

Rating Level 5

Count all medium helicopters (e.g., HIND, HIP, HAZE, HOUND, PUMA, WASP).

Detect deployed TWIN EAR antenna.

Distinguish between river crossing equipment and medium/heavy armored vehicles by size and shape (e.g., MTU-20 versus T-62 MBT).

Table 5.5 (cont.)

Detect missile support equipment at an SS-25 RTP (e.g., TEL, MSV).

Distinguish bow shape and length/width differences of SSNs.

Detect the break between railcars (count railcars).

Rating Level 6

Distinguish between variable and fixed-wing fighter aircraft (e.g., FENCER versus FLANKER).

Distinguish between the BAR LOCK and the SIDE NET antennas at a BAR LOCK/SIDE NET acquisition radar site.

Distinguish between small support vehicles (e.g., UAZ-69, UAZ-469) and tanks (e.g., T-72, T-80).

Identify SS-24 launch triplet at a known location.

Distinguish between the raised helicopter deck on a KRESTA II (CG) and the helicopter deck with main deck on a KRESTA I (CG).

Identify a vessel by class when singly deployed (e.g., YANKEE I, DELTA I, KRIVAK II FFG).

Detect cargo on a railroad flatcar or in a gondola.

Rating Level 7

Identify small fighter aircraft by type (e.g., FISHBED, FITTER, FLOGGER).

Distinguish between electronics van trailers (without tractor) and van trucks in garrison.

Distinguish by size and configuration between a turreted, tracked APC and a medium tank (e.g., BMP-1/2 versus T-64).

Detect a missile on the launcher in an SA-2 launch revetment.

Distinguish between bow-mounted missile system on KRIVAK i/II and bow-mounted gun turret on KRIVAK III.

Detect road/street lamps in an urban, residential, or military complex.

Rating Level 8

Distinguish the fuselage difference between a HIND and a HIP helicopter.

Distinguish between the FAN SONG E missile control radar and the FAN SONG F based on the number of parabolic dish antennas (three versus one).

Identify the SA-6 transloader when other SA-6 equipment is present.

Distinguish limber hole shape and configuration differences between DELTA I and YANKEE I (SSBNs).

Identify the dome/vent pattern on rail tank cars.

Rating Level 9

Detect major modifications to large aircraft (e.g., fairings, pods, winglets).

Identify the shape of antennas on EW/GCI/ACQ radars as parabolic, parabolic with clipped corners, or rectangular.

Identify, based on presence or absence of turret, size of gun tube, and chassis configuration, wheeled or tracked APCs by type (e.g., BTR-80, BMP-1/2, MT-LB, M113).

Identify the forward fins on an SA-3 missile.

Identify individual hatch covers of vertically launched SA-N-6 surface-to-air system.

Identify trucks as cab-over-engine or engine-in-front.

Table 5.6
Civil National Imagery Interpretability Rating Scale—March 1996

Rating Level 0

Interpretability of the imagery is precluded by obscuration, degradation, or very poor resolution.

Rating Level 1

Distinguish between major land use classes (e.g., urban, agricultural, forest, water, barren).

Detect a medium-sized port facility.

Distinguish between runways and taxiways at a large airfield.

Identify large area drainage patterns by type (e.g., dendritic, trellis, radial).

Rating Level 2

Identify large (i.e., greater than 160 acres) center-pivot irrigated fields during the growing season.

Detect large buildings (e.g., hospitals, factories).

Identify road patterns, like cloverleafs, on major highway systems.

Detect ice-breaker tracks.

Detect the wake from a large (e.g., greater than 300 ft) ship.

Rating Level 3

Detect large area (greater than 160 acres) contour plowing.

Detect individual houses in residential neighborhoods.

Detect trains or strings of standard rolling stock on railroad tracks (not individual cars).

Identify inland waterways navigable by barges.

Distinguish between natural forest stands and orchards.

Rating Level 4

Identify farm buildings as barns, silos, or residences.

Count unoccupied railroad tracks along right-of-way or in a railroad yard.

Detect basketball court, tennis court, volleyball court in urban areas.

Identify individual tracks, rail pairs, control towers, switching points in railyard.

Detect jeep rails through grassland.

Rating Level 5

Identify Christmas tree plantations.

Detect open bay doors of vehicle storage buildings.

Identify tents (larger than two person) at established recreational camping areas.

Table 5.6 (cont.)

Distinguish between stands of coniferous and deciduous trees during leaf-off condition.

Detect large animals (e.g., elephants, rhinoceros, giraffes) in grasslands.

Rating Level 6

Detect narcotics intercropping based on texture.

Distinguish between row (e.g., corn, soybean) crops and small grain (e.g., wheat, oats) crops.

Identify automobiles as sedans or station wagons.

Identify individual telephone/electric poles in residential neighborhoods.

Detect foot trail through barren neighborhoods.

Rating Level 7

Identify individual mature cotton plants in a known cotton field.

Identify individual railroad ties.

Detect individual steps on a stairway.

Detect stumps and rocks in forest clearings and meadows.

Rating Level 8

Count individual baby pigs.

Identify a USGS benchmark set in a paved surface.

Identify grill detailing and/or the license plate on a passenger/truck type vehicle.

Identify individual pine seedlings.

Identify individual water lilies on a pond.

Identify windshield wipers on a vehicle.

Rating Level 9

Identify individual grain heads on small grain (e.g., wheat, oats, barley).

Identify individual barbs on a barbed wire fence.

Detect individual spikes in railroad ties.

Identify individual bunches of pine needles.

Identify an ear tag on large animals (e.g., deer, elk, moose).

Table 5.7
Multispectral Imagery Interpretability Rating Scale—February 1995

Level 1

Distinguish between urban and rural areas.

Identify a large wetland (greater than 100 acres).

Detect meander flood plains (characterized by features such as channel scars, oxbow lakes, meander scrolls).

Delineate coastal shorelines.

Detect major highway and rail bridges over water (e.g., Golden Gate, Chesapeake Bay).

Delineate extent of snow or ice cover.

Level 2

Detect multilane highways.

Detect strip mining.

Determine water current direction as indicated by color differences (e.g., tributary entering large water feature, chlorophyll or sediment pattern).

Detect timber clear-cutting.

Delineate extent of cultivated land.

Identify riverine flood plains.

Level 3

Detect vegetation/soil moisture differences along a linear feature (suggesting the presence of a fence line).

Identify major street patterns in urban areas.

Identify golf courses.

Identify shoreline indications of predominant water currents.

Distinguish among residential, commercial, and industrial areas within an urban area.

Detect reservoir depletion.

Level 4

Detect recently constructed weapons positions (e.g., tank, artillery, self-propelled gun) based on the presence of revetments, berms, and ground scarring in vegetated areas.

Distinguish between two-lane improved and unimproved roads.

Detect indications of natural surface airstrip maintenance or improvements (e.g., runway extensions, grading, resurfacing, brush removal, vegetation cutting).

Detect landslide or rockslide large enough to obstruct a single-lane road.

Detect small boats (15–20 feet in length) in open water.

Identify small areas suitable for use as light fixed-wing (e.g., Cessna, Piper Cub, Beechcraft) landing strips.

Table 5.7 (cont.)

Level 5

Detect automobile in a parking lot.

Identify beach terrain suitable for amphibious landing operation.

Detect ditch irrigation of beet fields.

Detect disruptive use of paints or coatings on buildings/structures at a ground forces installation.

Detect raw construction materials in ground forces deployment areas (e.g., timber, sand, gravel).

Level 6

Detect summer woodland camouflage netting large enough to cover a tank against a scattered tree background.

Detect foot trail through tall grass.

Detect navigational channel markers and mooring buoys in water.

Detect livestock in open but fenced areas.

Detect recently installed minefields in ground forces deployment areas based on a regular pattern of disturbed earth or vegetation.

Count individual dwelling in subsistence housing areas (e.g., squatter settlements, refugee camps).

Level 7

Distinguish between tanks and three-dimensional tank decoys.

Identify individual 55-gallon drums.

Detect small marine mammals (e.g., harbor seals) on sand/gravel beaches.

Detect underwater pier footings.

Detect foxholes by ring of spoil outlining hole.

Distinguish individual rows of truck crops.

With interval scaling, raters are allowed, and even encouraged, to use ratings outside of the 0 to 100 continuum. The results of the image scaling evaluation are analyzed to define image marker chips equally spaced (quartiles or quintiles) across a 0 to 100 continuum. Next, a set of interpretation tasks or criteria is assembled for each of the criteria categories. Each criterion consists of three parts, a recognition level, an object type, and a descriptive qualifier. Figure 5.13 provides examples. IAs then rate the criteria against the image SQS in terms of the image scale level needed to accomplish the exploitation task. For example, if the IA believes the task can be accomplished on the 60 marker but not the 40, the task would receive a rating between 40 and 60 with the final value representing the IA's judgement as to the relative level of interpretability required. The resultant data is analyzed to define criteria

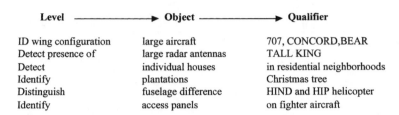

Level ⟶	Object ⟶	Qualifier
ID wing configuration	large aircraft	707, CONCORD,BEAR
Detect presence of	large radar antennas	TALL KING
Detect	individual houses	in residential neighborhoods
Identify	plantations	Christmas tree
Distinguish	fuselage difference	HIND and HIP helicopter
Identify	access panels	on fighter aircraft

Figure 5.13 NIIRS criteria examples.

equally spaced across the rating continuum and exhibiting minimum rating variance. Assuming a 0–100 SQS continuum, level 1 of the NIIRS would be at an SQS rating of 11 and level 9 at 99. This is done for each of the criteria categories after normalization across the criteria categories to achieve category equivalence. The selected criteria define the scale.

The final step in the development process is a validation study. A sample of imagery is rated by IAs and the data analyzed to determine that the scale is linear, has equal intervals, is system and criteria category independent,[5] and is understandable to the IAs. Data from the development process is used to develop and publish NIIRS manuals and training and certification materials. IAs are required to pass a rating test before they are certified to make NIIRS ratings.

Three types of NIIRS ratings are made [17]. Absolute NIIRS ratings use the criteria to rate imagery. Absolute NIIRS ratings can be made at the decimal level by experienced IAs. Absolute NIIRS standard deviations typically range from 0.6 to 0.8. A rating can also be made on the NIIRS difference between two images. This is termed a delta NIIRS rating, and standard deviations range from 0.2 to 0.3 (assuming the true difference is less than one NIIRS). Finally, calibrated image sets have been developed called engineering NIIRS standard (ENS) sets. Images of known NIIRS ratings make up a set and are labeled from A to G or H in order of increasing quality. The NIIRS level is unknown to the rater, who places the image to be rated on the ENS continuum. Adjacent ENS images are separated by a 10-point scale. Average ENS standard deviations are on the order of 0.4. Except at the very extremes of the scale, which are not encountered in most studies, rating variability is not correlated highly with rating means. Variance is thus homogeneous. Figure 5.14 shows a set of data taken from a video compression study [21]. The value of R^2 indicates the proportion of variance in Y accounted for by variations in X (mean NIIRS). R^2 is the square of the correlation coefficient. In this case, there is no statistically significant relation-

---

5. Ratings for the same quality level do not vary as a function of system or criteria category.

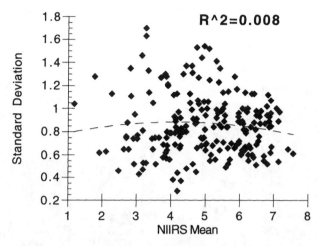

Figure 5.14 NIIRS means versus variability. (Data from [22].)

ship. NIIRS is, by definition, linear with respect to physical image quality. Because rating variability is quite low, reliability is generally high. In the same study where probability of identification data was acquired and shown in Figure 5.3, NIIRS ratings were also made on the same stimuli. The NIIRS reliability was 0.98 (compared to 0.88 for the probability data).

5.4.2 Task Satisfaction Confidence Ratings

The nine NIIRS criteria selected for each criteria category or order of battle are selected from over 100 tasks used in the scale development process. These in turn represent only a small sample of the tasks an IA might be asked to perform. Furthermore, in the desire to reduce rating variability, the NIIRS criteria are very specific. For example, visible NIIRS 5 requires the ability to distinguish between (identify) two specific aircraft models. The level required to identify other models is unspecified. For this reason, it is sometimes desirable to use TSC ratings. The task can be made as general or as specific as desired and there is no limit (other than time and the patience of participants) as to the number of tasks that can be addressed. The response is made on a 0–100 scale and indicates the analyst's confidence that the task can be performed on an image of the quality being viewed. If tasks are broadly stated (detect all tanks), response can be generalized (at the expense of response variability). Such data is more easily related to direct measures of performance. Tasks can also be specifically linked to information requirements. As an example, the National Imagery and Mapping Agency maintains a feature and attribute database [23]. Features are the types of information (e.g., roads,

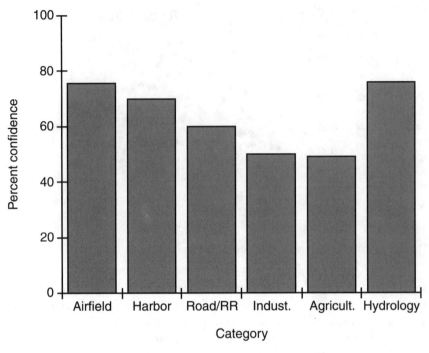

Figure 5.15 RADARSAT feature extraction performance. (Data from [24].)

railroads, streams) plotted on maps, and attributes are the characteristics of those features (e.g., four-lane divided highway). In an assessment of RADARSAT image utility [24], IAs were asked to assess their confidence in detecting and attributing selected features. Figure 5.15 provides a sample of results. This data was used to assess the utility of RADARSAT imagery for mapping support.

TSC ratings show some of the same statistical properties as do direct performance measures such as probability of detection. Ratings are nonlinear with respect to physical image quality [25]. Rating variability is correlated with the mean rating. Figure 5.16 provides an example. Here, there is a significant relationship between mean TSC ratings and variability. The transforms used for direct performance measurement data can be used with TSC rating data. Reliability ratings for TSC data are also similar to those for direct performance measures. A comparison of NIIRS and TSC reliabilities showed a value of 0.82 for TSC ratings and 0.93 for NIIRS [25].

Earlier, it was stated that IAs could accurately predict their performance. This has been demonstrated for positive responses (detection, identification) but not for negative responses (false alarms, misidentifications). A series of

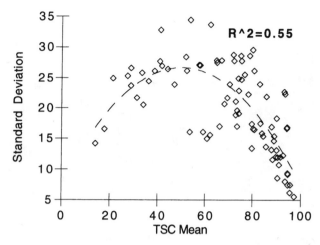

Figure 5.16 TSC means versus variability. (Data from [25].)

TSD studies dealing with detection of targets in synthetic radar imagery displayed in softcopy were performed [13]. In three of the studies, IAs provided detection responses along with confidence ratings. In three others (using the same data), they provided estimates of detection and false alarm performance. Confidence ratings provided with measured detection responses (self-assessment) showed an average correlation of 0.87 between confidence and probability of detection performance. The correlation between predictive confidence estimates and measured performance (prediction for others) was 0.80. Both values were statistically significant. Correlations between nontarget confidence ratings and performance were not statistically significant.

The validity of the predictions was assessed by determining the proportion of predictive estimates that correctly predicted measured performance. Figure 5.17 shows results for detections and false alarms. In this study, IAs were able to accurately predict their own detection performance, as well as that of others (although with slightly less accuracy). They were much less accurate in predicting false alarms.

In discussing TSD, it was stated that measures of detection and recognition performance could fall at any point along the ROC, depending on a variety of factors affecting the observer. It is not clear that this is true to the same degree with estimated measures of performance. The previously cited data, for example, accurately predicted detection performance at a $\beta = 1$ level. The instructions were designed to maintain this level by explicitly stating that there would be an equal number of target and nontarget presentations and that the cost of a missed detection was the same as the cost of a false alarm. In most TSC studies, this level of instruction is not provided.

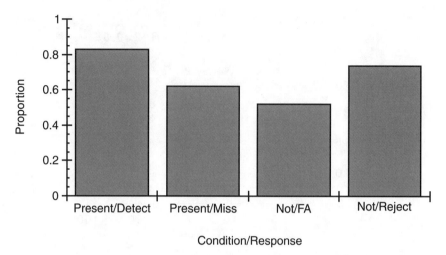

Figure 5.17 Confidence ratings versus measured performance. (Data from [13], © 2000, Veridian Systems/Veridian ERIM International, Inc.)

In a study of the use of confidence ratings to determine whether or not an IA's response should be checked by another IA [26], a strong positive relationship was found between identification accuracy and confidence ratings. There was no relationship, however, between confidence and a determination that all of the targets present had been found (confidence versus misses). It thus appears that IAs can accurately predict positive performance, but not negative performance. The tendency appears to be one of underestimating errors.

5.4.3 Relative Quality Scales

The final category of performance estimate scales are termed "relative quality" scales. They are scales of relative quality and do not relate directly to information extraction performance, although they typically show a high correlation. The subjective image and criteria scaling used in the development of the NIIRS is an example of subjective or relative quality scaling. Another common scale relates image quality deficiencies (e.g., noise, lack of sharpness) to their impact on image interpretability using a 10-point scale (Figure 5.18). All of the commonly used psychophysical scaling approaches have found their way into studies of image quality and utility, ranging from simple paired comparison to multidimensional, categorical, interval, and ratio scaling.

Studies of commercial video have relied almost totally on relative quality scales. Current video models use such ratings and consequently they will be

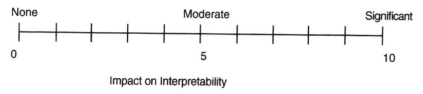

Figure 5.18 Ten-point interpretability impact scale.

discussed in some detail. The two most common are the Double Stimulus Continuous Quality Scale (DSCQS) and the Double Stimulus Impairment Scale (DSIS). In both cases, a standard method of data collection and analysis has been defined and implemented. This includes display specification, observer screening, and data analysis and reporting procedures. These procedures are defined in Recommendation ITU-R BT.500-10 [27].

DSCQS ratings are made on a graphic five-point scale. Figure 5.19 shows an example. Two video sequences are shown for 10 seconds each, with a 3-second separation. One of the sequences is the original and the other a processed version (e.g., bandwidth compression). The processed version is labeled B and the original A. The sequences can be repeated until the observer has formed a judgment. The rating is then made for each of the versions. The graphic responses are converted to 0–100 (sometimes 0–5) values, and analysis is performed. The differences between the two ratings are used. Another method of analysis is to define a rating for B as a percentage of the rating of A [28]. Figure 5.20 provides an example.

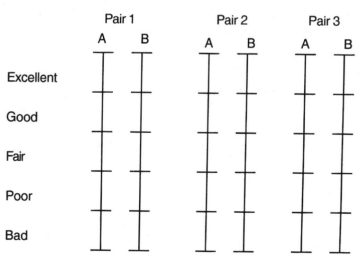

Figure 5.19 DSCQS rating form example.

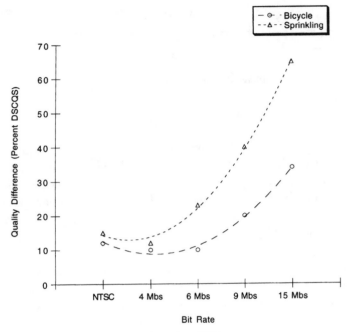

Figure 5.20 DSCQS rating data example. (Data from [28].)

DSIS ratings are used to assess the impact of potential quality degradation factors. As with DSCQS ratings, the observer views pairs of video clips. One is the reference and the second is the sequence to be rated. Unlike DSCQS ratings, only the potentially impaired sequence is rated using the scale shown in Figure 5.21. The rating is made by assigning an impairment rating to the clip using the scale shown in the figure.

Sequence	Response	Scale	
1	☐	5	imperceptible
		4	perceptible, but not annoying
2	☐	3	slightly annoying
		2	annoying
3	☐	1	very annoying
4	☐		

Figure 5.21 DSIS rating form example.

Relative quality scales tend to show a correlation between mean ratings and variability. Figure 5.22 shows an example using data from a study of video data compression [25] where analysts rated the impact (on interpretability) of artifacts resulting from bandwidth compression. As is the case with TSC ratings, there is a statistically significant relationship between mean ratings and variability. Similar results have been shown for DSCQS ratings. Because of so-called end-effects, relative quality ratings tend to be nonlinear with respect to physical image quality. There is often a tendency to avoid using extremes of the scale. For this reason, images at the extremes of the quality scale tend to receive lower (good quality) or higher (poor quality) ratings than would otherwise be the case. It is for this reason that subjective quality ratings made in the NIIRS development process are allowed to exceed the 0 and 100 ratings.

Perhaps because there is no objective standard, relative quality scales tend to show greater rating variability than, say, NIIRS or even TSC rating scales. Figure 5.23 shows data from two studies of video compression [22,25] employing NIIRS, TSC, relative quality, and DSCQS ratings [28,29]. Variability is expressed as average rating variability (standard deviation) as a percentage of total scale. There was a strong correlation between artifact rating (AR) means and standard deviations (0.82). The AR scale is a 10-point scale where 0 indicates no effect and 10 indicates a significant effect. Reliability was 0.85. In the study using the DSCQS, rating differences were defined

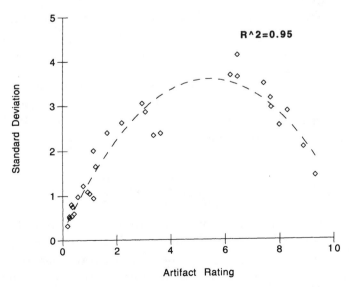

Figure 5.22 Relationship of mean artifact rating and variability. (Data from [25].)

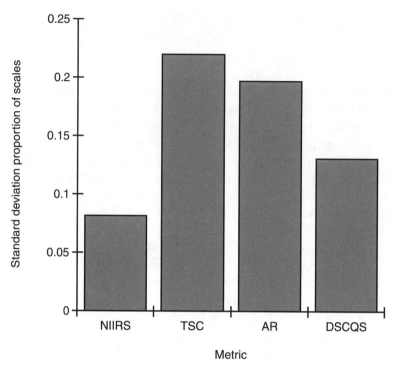

Figure 5.23 Comparison of rating variability across relative quality measures. (Data from [25].)

between the original video clip and sequences processed through a variety of hypothetical reference circuits (HRCs) of varying bit rates. The average standard deviation of the difference score was on the order of 10–15 units (of 100) and there was no obvious relationship between difference scores and score variability [25].

As was the case with the relationship between direct probability measures and TSC estimates, measurements made on the same imagery tend to be highly correlated. Figure 5.24 shows the correlation between NIIRS, TSC, and ratings on a 10-point relative quality scale [22].

5.5 Information and Difference Metrics

In this final grouping of performance prediction metrics, we include both information and difference metrics. In a practical sense, these metrics fall between the true performance prediction (dependent) variables we have discussed previously in this chapter and the performance predictor (independent) variables discussed in Chapter 6. Mathematically, they are dependent

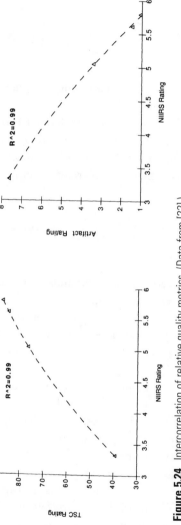

Figure 5.24 Intercorrelation of relative quality metrics. (Data from [22].)

variables in that they vary as a function of physical image quality variables. They are more often used, however, as an intermediate step in predicting the values of perceptual metrics such as NIIRS and DSCQS ratings.

In Chapter 1, the concept of information density was introduced. It was defined as

$$ID = \pi \int_0^\infty \log_2 \left(1 + \frac{P_S(r)}{P_N(r)} \right) r \, dr \qquad (5.10)$$

where $P_S(r)$ = signal power for spatial frequency r, and $P_N(r)$ = noise power for spatial frequency r. Frequency is defined in polar coordinates. For sampled imagery, the upper limit of integration was defined as the Nyquist limit. The function assumes a radially symmetric FFT image and results in a one-dimensional plot of information density. Information can also be defined as a function of spatial frequency [30]. Here, it is defined as

$$ID(r) = \log_2 \left(1 + \frac{P_S(r)}{P_N(r)} \right) \quad \text{[bits/cycle]} \qquad (5.11)$$

Where quantizing error occurs, the noise spectrum must be modified to account for the different effect of quantizing error on the noise spectrum. The formulation in (5.11) requires that noise be Gaussian, additive in amplitude, and independent of signal level. Quantizing noise is related to the width of the quantizing interval and is more rectangular than Gaussian in distribution.

Typical metrics used to predict direct or predicted performance plot information density as a function of spatial frequency, perform some type of normalization function, typically reduce the 2-D spectrum to a 1-D plot, and sometimes account for human visual performance limitations. The integrated area under some portion of the plot, with various adjustments for such things as image scale, is defined as a measure of information. The measure is a single value based on a corrected area measure of bits (information density) per unit area. The relationship between this measure and the interpretability or performance metric is then empirically defined. The relationship between an image quality metric (IQM) based on the digital power spectrum has been shown to be linear with respect to NIIRS [1]. The log of information density based on the optically derived power spectrum has been shown to be nonlinear with respect to recognition accuracy. Figure 5.25

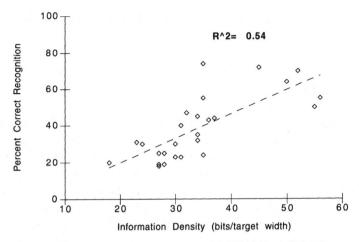

Figure 5.25 Correlation of information density and recognition. (Data from [30].)

shows data from [30]. The power spectrum approach is generally based on the assumption that the input scene power spectrum is constant for all scenes. This assumption is only approximately correct. Power spectrum–derived measures have successfully been used for scene content classification [31,32], and substantial variations can occur within a scene based on content [1]. Despite this, global scene measures have been shown to be sensitive to variations in image quality. Because information density is, in theory, infinite, there is no reason to believe that there is a correlation between measurement means and standard deviations due to an "end-effect." An end-effect occurs when a scale is limited in magnitude and all predictors or estimates of scale values converge to a single value. This was previously shown to occur with TSC and DSCQS ratings where at the ends of the scale all observers (or predictors) agree on the rating.

Difference metrics are relative measures in that they are scene content variant. They can thus be used only to compare different versions of the same scene. Perhaps the most common metric is the just-noticeable difference (JND) technique based on models of the human visual system. The JND metric is a measure of the difference between two versions of the same scene. The Sarnoff JND model [2] and the digital video quality (DVQ) metric [33] are examples. Because the metric is always used as a difference measure, there is no good measure of variability. The relationship between JND values and relative quality as measured by DSCQS ratings was shown to be nonlinear [34]. Although not specifically tested, there is a suggestion that the relationship between the Sarnoff JND metric and relative 0–100 quality ratings, as well as NIIRS, is nonlinear. Since JNDs are not limited in magnitude, there is no reason to believe there is a relationship between means and variance due to the end-effect.

5.6 Summary

Three broad classes of performance variables were discussed. Direct performance measures are calculated from actual observer performance and are typically expressed as probability values. They have the greatest face validity, but require substantial data collection resources. They are also subject to substantial variability because of variations in the observer's decision criterion. TSD is a method for examining and quantifying variations in the observer's decision criterion. Over a wide range of image quality, direct performance measures tend to show a logit function relationship (asymptotic at high and low levels of quality). Other direct measures include accuracy and completeness (derived from detection and identification data) and time or distance to achieve defined levels of detection/identification performance.

Three types of performance estimate measures were reviewed. NIIRS is a rating scale defining the level of interpretability (information extraction) that can be achieved on a given image. TSC ratings are the predictive correlate to directly measured detection and identification performance. TSC ratings accurately predict positive performance (detection and identification) but not negative performance (misses, false alarms). Finally, a variety of relative quality scales are used to define relative performance or quality. DSCQS ratings are perhaps the best documented of this type.

A final class of information and difference metrics was briefly presented. Information metrics are derived from the image power spectrum and are, in turn, used to predict performance estimate measures such as NIIRS or directly measured probabilities of performance. Difference metrics such as JND estimates are derived from alternative processing of the same scene image and are also used to predict performance.

References

[1] Nill, N. B., and B. H. Bouzas, "Objective Image Quality Measure Derived from Digital Image Power Spectra," *Optical Engineering*, Vol. 31, No. 4, 1992.

[2] Lubin, J., *A Methodology for Imaging System Design and Evaluation*, Princeton, NJ: David Sarnoff Research Center, 1995.

[3] Hodges, J. S., and J. A., Dewar, *Is It You or Your Model Talking? A Framework for Model Validation*, R-4114-AF/A/OSD, Santa Monica, CA: Rand Corporation, 1992.

[4]   Beechler, R.L., et al., *A Study of Rapid Photointerpretation Methods,* TRR 1153, Washington, DC: U.S. Army Behavioral Research Laboratory, 1969.

[5]   Driggers, R., et al., "Sensor Performance Conversions for Infrared Target Acquisition and Intelligence-Surveillance-Reconnaissance [ISR] Imaging Sensors," *Applied Optics,* Vol. 38, No. 28, 1999, pp. 5936–5943.

[6]   *StatView Reference Manual,* Cary, NC: SAS Institute, 1998.

[7]   Winer, B. J., *Statistical Principles in Experimental Design,* New York, NY: McGraw Hill, 1962.

[8]   Green, D. M., *Signal Detection Theory and Psychophysics,* Huntington, NY: Robert Krieger Publishing Company, 1974.

[9]   Swets, J. A., (ed.), *Signal Detection and Recognition by Human Observers,* New York, NY: Wiley, 1964.

[10] Egan, J. P., *Signal Detection Theory and ROC Analysis,* New York, NY: Academic Press Inc., 1975.

[11] Irvine, J., *Estimation of Free Response Operating Characteristic Curves,* Meetings of the American Statistical Association, 1990.

[12] Bunch, P. C., et al., "A Free-Response Approach to the Measurement and Characterization of Radiographic Observer Performance," *Journal of Applied Photographic Engineering,* Vol. 4, No. 4, 1978.

[13] Voas, R. B., W. Frizzell, and W. Zink, *Evaluation of Methods for Studies Using the Psychophysical Theory of Signal Detection,* Washington, DC: Environmental Research Institute of Michigan, 1987.

[14] Jennings, L. B., et al., *Ground Resolution Study Final Report,* RADC-TR-63-224, Griffiss Air Force Base, NY: Rome Air Development Center, 1963.

[15] Bennett, C. A., et al., *A Study of Image Qualities and Speeded Intrinsic Target Recognition,* Owego, NY: IBM Federal Systems Division, 1963.

[16] Birnbaum, A. H., *Effect of Selected Photo Characteristics on Detection and Identification,* Washington, DC: U.S. Army Personnel Research Office, 1962.

[17] Leachtenauer, J. C., "National Image Interpretability Rating Scales: Overview and Product Description," *ASPRS/ASCM Annual Convention & Exhibition,* Vol. 1, Baltimore, Maryland, April 1996, pp. 262–271.

[18] Hothem, D., et al., "Quantifying Imagery Interpretability for Civil Users," *ASPRS/ASCM Annual Convention & Exhibition,* Vol. 1, Baltimore, Maryland, April 1996, pp. 292–298.

[19] Greer, J. D., and J. Caylor, "Development of and Environmental Image Interpretability Rating Scale," *SPIE,* Vol. 1763, 1992, pp. 151–157.

[20] Mohr, E., et al., "The Multispectral Imagery Interpretability Rating Scale (MS IIRS)," *ASPRS/ASCM Annual Convention & Exhibition,* Vol. 1, Baltimore, Maryland, April 1996, pp. 300–310.

[21] Irvine, J. M., and J. C. Leachtenauer, "A Methodology for Developing Image Inter-pretability Scales," *ASPRS/ASCM Annual Convention & Exhibition*, Vol. 1, Baltimore, Maryland, April 1996, pp. 273–279.

[22] Leachtenauer, J., M. Richardson, and P. Garvin, "Video Data Compression Using MPEG-2 and Frame Decimation," *Proceedings of SPIE, Visual Information Processing VIII*, Vol. 3716, Orlando, Florida, April 6, 1999, pp. 42–51.

[23] Defense Mapping Agency, *DMA Standard Supporting Mark-90, Section 100, Glossary of Feature Attribute Definitions*, PS/7BC/000/199310, Fourth Edition, May 1994.

[24] Vrabel, J., and J. Leachtenauer, "Civil and Commercial Application Project for Satellite Imaging Systems," *ASPRS Annual Convention*, Tampa Bay, Florida, April 1998, pp. 150–162.

[25] Leachtenauer, J. C., "Comparison of Video Compression Evaluation Metrics for Military Applications," *Proceedings of SPIE*, Vol. 3959, January 2000, pp. 88–98.

[26] Samet, M. G., *Checker Confidence as Affected by Performance of Initial Image Interpreter*, Technical Research Note 214, Washington, DC: U.S. Army Behavioral Science Research Laboratory, 1969.

[27] ITU-R Recommendation BT 500-10, *Methodology for the Subjective Assessment of the Quality of Television Pictures*, International Telecommunication Union Recommen-dation, 1974–2000.

[28] Nakasu, E., et al., "A Statistical Analysis of MPEG-2 Picture Quality for Television Broadcasting," *SMPTE Journal*, November 1996, pp. 702–711.

[29] Jones, C., et al., *Analysis of T1A1.5 Subjective and Objective Test Data*, Document Number T1A1.5/94-152, National Telecommunications and Information Adminis-tration, Institute for Telecommunication Sciences, October 1994.

[30] Schindler, R. A., *Optical Power Spectrum Analysis of Display Imagery*, AMRL-TR-78-50, Wright-Patterson Air Force Base, OH: Aerospace Medical Research Laboratory, 1978.

[31] Gramenopolous, N., "Automated Thematic Mapping and Change Detection of ERTS-1 Images," *Proceedings, Conference on Management and Utilization of Remote Sensing Data*, Sioux Falls, South Dakota, 1973, pp. 432–446.

[32] Leachtenauer, J. C., "Optical Power Spectrum Analysis," *Photogrammetric Engineering and Remote Sensing*, Vol. 43, No. 9, 1977, pp. 117–1125.

[33] Watson, A. B., "Toward a Perceptual Video Quality Metric," *IS&T/SPIE Conference on Human Vision and Electronic Imaging III*, SPIE, Vol. 3299, San Jose, California, 1998, pp. 139–147.

[34] Martens, J.-B., and L. Meesters, "The Role of Image Dissimilarity in Image Quality Models," *IS&T/SPIE Conference on Human Vision and Electronic Imaging IV*, SPIE, Vol. 3644, January 1999, pp. 258–269.

6

Information Extraction Performance Predictors

In the previous chapter, measures of performance used to characterize the information extracted from S&R systems were discussed. In this chapter, the variables used in predicting those measures are introduced. Some of these were discussed in Chapter 2 as a means of characterizing S&R systems. In Chapter 4, the impulse response function, or point spread function (PSF), and its Fourier transform, the modulation transfer function (MTF), were introduced as a means of describing the performance of LSI systems. Here, we step back and approach such metrics in the context of the observer performance prediction literature. The metrics range from simple and easily defined measures such as scale and number of scan lines to more complex measures such as those based on the modulation transfer function. The measures are defined, and examples of performance data are shown.

Image quality has been defined [1] as the extent to which a displayed image duplicates the information contained in the imagery in a form suitable for viewing by an imagery analyst (IA). An image is a representation of a scene or object. Schott [2] makes a distinction between quality and fidelity. Fidelity is defined by Schott as the extent to which a remote sensing system reproduces the characteristics of interest (relative to a scene or object). Schott notes that quality takes account of the visual and psychophysical characteristics of the observer. Because of this, high quality is not always high fidelity.

Fidelity (in terms of an image) can be defined along three dimensions: spatial, radiometric, and geometric or geospatial. Spatial fidelity is the extent to which the image preserves the relative size, shape, and detail of the object. Radiometric fidelity is the extent to which the image preserves the relative or absolute energy distribution in the scene. Geometric or geospatial fidelity is the extent to which the image preserves relative or absolute positions within a scene. Image quality can be defined along the same dimensions, although not necessarily with a 1:1 mapping. For example, contrast enhancement may result in image quality improvement at the expense of radiometric fidelity.

In modeling the performance of S&R systems, it is generally image quality, rather than image fidelity, that is of concern. Geometric quality or fidelity is also seldom of concern because the human visual system is quite adaptable to dealing with geometric distortion. So long as relative relationships are reasonable, performance is not affected. The exceptions are those cases where physical size or distance measurements are required. Here, any random distortions are of concern because they can not be compensated for in the measurement process. Most S&R systems exhibit some type of distortion, but these distortions are generally ignored from a performance prediction viewpoint. Radiometric fidelity is also generally of little concern for visual interpretation but is an issue with machine processing algorithms.

Although there is general agreement that image resolution, contrast, and noise are key factors in defining S&R system performance, there are variations in how these factors are measured. Other factors such as image size, artifacts, and viewing conditions also affect performance. Much effort has been expended on developing summary measures of image quality. The modulation transfer function area (MTFA) is an example of a summary measure [3], as are image power spectrum approaches [4]. In the discussions that follow, EO, IR, and SAR metrics and data are described. Because many of the metrics used with EO systems were originally developed with photographic imagery, photographic metrics are also described.

6.1   Scale, Resolution, and Sharpness

Scale, resolution, and sharpness all relate to the size of the smallest detail in an image that can be seen. In a very broad sense, the system user or designer has an indication of the size of the targets or target detail of interest. Scale and resolution are thus intuitive (although not very accurate) measures of system performance.

6.1.1   Scale

Until sometime in the 1950s, image scale was the variable used by the reconnaissance community to task systems. For a camera, scale (S) is defined as

$$S = f/H \qquad (6.1)$$

where H = altitude in feet, and f = camera focal length in feet. It is also defined more generally for any system by the ratio of image to ground distance:

$$S = ID/GD \qquad (6.2)$$

where ID = image distance and GD = ground distance, both in common units and the ratio normalized to 1 (e.g., 1/250,000, 1/24,000). The concept is illustrated in Figure 6.1. Because image scale is expressed as a fraction, a smaller denominator in the fraction denotes larger scale. As the denominator increases, scale becomes smaller. The term "scale factor," sometimes written as SF, is the inverse of image scale. An image with a scale of 1/18,000 has a scale factor of 18,000.

For a given system, scale defines the size relationship between ground and image distance. It is thus a measure of what can and cannot be seen on the image, all other factors being equal. Figure 6.2 provides an example.

The 1957 Joint Technical Manual of Photo Interpretation [5] provides a table showing the photographic scales required to accomplish a variety of tasks. Table 6.1 provides a partial example. The table was intended as a guide for personnel tasked with planning photo reconnaissance missions. The guidance provided with the table is illuminating: "These minimum scales are the educated conclusions of a qualified panel of photo interpreters. It is not expected that they will achieve concurrence with the opinions or experience of all who refer to them. The many variables in atmospheric conditions and film processing to which an aerial photo is exposed will tend to alter these scales.

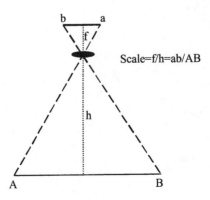

Figure 6.1 Image scale.

1/3750   1/7500

Figure 6.2 Effects of image scale.

These scales are based on the average quality photography currently being produced."

6.1.2 Photographic Resolution

The same table printed in the 1957 Joint Technical Manual of Photo Interpretation, with some additions but without the qualifying explanation, was

Table 6.1

Scale Requirements

Target	Minimum scale: Identification	Minimum scale: Technical analysis
Basic industry	1/30,000	1/12,000
Heavy anti-aircraft guns	1/15,000	1/3000 (low-level oblique)
Light anti-aircraft guns	1/2000	1/2000 (low-level oblique)
Motor vehicles	1/10,000	1/2000
Rail	1/30,000	1/8000
Road	1/30,000	1/5000
Ships larger than DD*	1/25,000	1/12,000
DD* and minor combatants	1/15,000	1/7000
Sub	1/25,000	1/5000 (low-level oblique)
Fixed radar	1/10,000	1/5000
Mobile radar	1/8000	1/5000
Aircraft, ws < 40 ft	1/10,000	1/2000
Aircraft, ws > 100 ft	1/20,000	1/5000

*Destroyer.

published as an updated technical manual in 1967 [6]. A caveat was provided that indicated that the scale values "were computed for cameras having an average system resolution of 15–20 lines per millimeter. Photographs from cameras known to have higher or lower resolution values should be adjusted accordingly." It was thus recognized that variations in system resolution would affect the level of information extraction. "Resolution" is a general term used to evaluate the spatial performance of a system. For film return systems, the term "resolving power" is sometimes used. Resolving power is the number of line pairs per unit distance (commonly millimeters) that can be resolved on film. Figure 6.3 shows an example of a tri-bar resolution target. The standard Air Force (1951) tri-bar target shows patterns differing in size by $\sqrt[6]{2}$ [7]. This equates to a 12% change in spatial frequency between adjacent patterns. A pattern is said to be resolved if an observer can distinguish the individual lines in the pattern. If the lines of the smallest resolvable pattern are 0.1 mm in width, the resolving power is said to be five line pairs $(1/(0.1 \times 2))$ per millimeter.

The resolving power of a system is typically measured on an optical bench in the laboratory. This measurement accounts for the performance of the optics and film but not all of the other elements of the image chain such as camera vibration and atmospheric effects.

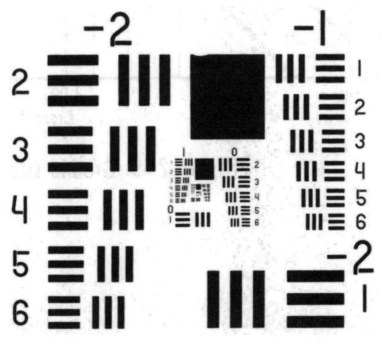

Figure 6.3 Tri-bar target.

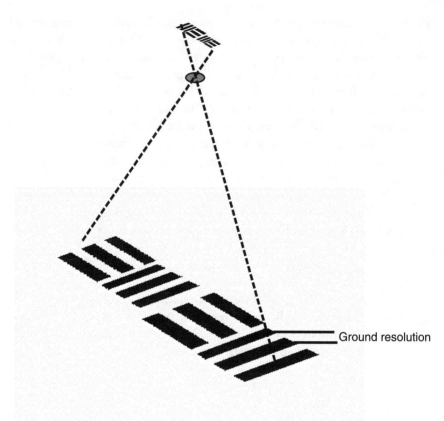

Figure 6.4 Ground resolution.

To account for the full image chain, the concept of ground resolution was introduced. Ground resolution is the size of the smallest pattern on the ground that can be resolved (on the image of the ground). It is the width (typically in feet) of the smallest resolvable bar and space (Figure 6.4).

Ground resolution is defined as [8]

$$GR = \frac{h}{(f)(R_{LF})25.4} \qquad (6.3)$$

where GR is ground resolution in feet, h is altitude in feet, f is focal length in inches, R_{LF} is system resolution in line pairs per millimeter, and 25.4 is a conversion from millimeters to inches. The term "ground-resolved distance"

(GRD) is also commonly used and has the same meaning as ground resolution.

Ground resolution can be measured in one of several ways. A resolution target on the ground can be imaged and read out on the resultant image. A resolution target can be imaged in the laboratory (test stand) and the resolution of the image extrapolated using an assumed scale factor. A visual edge matching (VEM) technique can be used [9]. Here, an edge in the image is matched to one of a series of edges differing in contrast and resolution. Ground resolution can be defined as the intersection of the MTF and threshold detectability curve (see Section 6.4). A technique called visual image evaluation (VIE) has also been used [10]. The intercorrelations among the different methods of measuring ground resolution range from 0.83 to 0.95 [8]. Measurement variability (one standard deviation) is on the order of 12% [10].

As a predictor of performance, ground resolution has limitations. A major area of controversy is the contrast of the target used to measure resolution. When a tri-bar target is imaged, contrast is lost as a function of spatial frequency. One measure of contrast is contrast modulation, defined as

$$C_m = \frac{L - D}{L + D} \tag{6.4}$$

where L = brightness of light bar, and D = brightness of dark bar. As the resolution bars and spaces get smaller, contrast is reduced as diagrammed in Figure 6.5. Ultimately, there is no contrast between light and dark bars and the target cannot be resolved. A plot of modulation over a spatial frequency range defines the MTF. MTF is discussed in more detail in Section 6.4 and was also described in Chapter 4 where it was calculated.

A problem with resolution as a measure of system performance is that the greater the original contrast of the resolution target, the finer the measured resolution of the system. A 1000:1 contrast ratio target (light bars 1000 times brighter than dark bars) will provide finer resolution readings than will a 1.6:1 contrast ratio target. The latter ratio more nearly represents the typical contrast in a photographic scene [11]. There can thus be substantial ambiguity in a ground resolution value for a system.

A second issue concerning resolution is edge sharpness. It was shown by Higgins, et al. [12] that images with the same resolution could differ in terms of the sharpness of edges and that observers preferred the sharper images. A measure called "acutance" was derived to measure sharpness. It was

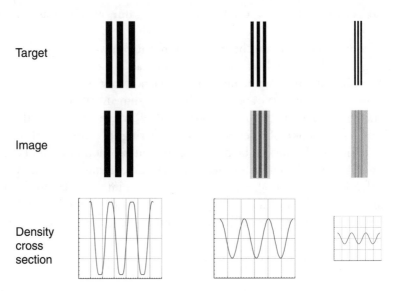

Figure 6.5 Effect of spatial frequency on contrast modulation.

based on the mean squared edge gradient evaluated by a microdensitometer trace across an exposed knife edge. The value G^2 is defined as

$$\overline{G_x^2} = \frac{\left[\Sigma_{i=1}^{n}(\Delta D_i / \Delta x_i)^2\right]}{n} \quad (6.5)$$

where $\overline{G^2}$ is the squared edge gradient, ΔD is the film density difference over distance x_i, and n is the number of positions. Acutance is defined as

$$A = \frac{\overline{G_x^2}}{D} \quad (6.6)$$

where $\overline{G_x^2}$ is as defined in (6.5) and D is the density range over which the measurements were made.

The effect of varying acutance is shown in Figure 6.6. Detection performance was measured on a set of images where acutance was degraded but resolution held constant [13]. Figure 6.6 shows a sample of results.

As contrast decreases, the effects of acutance are less. With low-contrast resolution targets, differences in acutance are of less importance [14].

A third issue in measuring resolution is the nature of the resolution target itself. Although tri-bar targets became the USAF standard, a variety

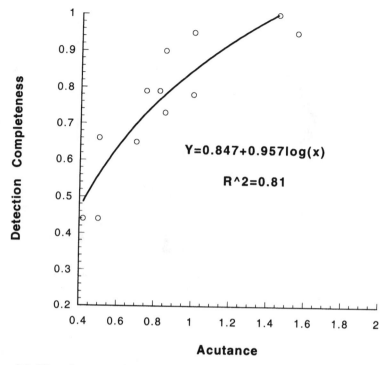

Figure 6.6 Effect of acutance changes in airfield detection. (Data from [13].)

of other targets have been used. Figure 6.7 provides examples. Results, in terms of resolving power for a given system and imaging operation, differ as a function of the target used. Figure 6.8 provides an indication of the differences under the best viewing conditions evaluated (background luminance of 500 fL). Values shown within the bars indicate target detail size and overall target size. The grating shows a substantially lower threshold than the Cobb target, even though both show the same bar separation.

6.1.3 Video Resolution (Analog)

The increased use of video cameras in the 1960s required a method of measuring resolution. Analog U.S. video is defined by SMPTE 170 M [15]. A frame consists of 525 lines in the vertical dimension. The 525 lines are defined in two fields, each field capturing every other line. A frame is thus two fields. Fields are displayed at ~60 Hz (actually 59.9 frames per second). Of the 525 lines, 483 to 490 are active, the remainder are used for vertical retrace (time required to move from bottom right to top left). The horizontal

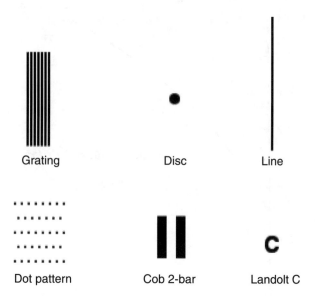

Figure 6.7 Resolution targets.

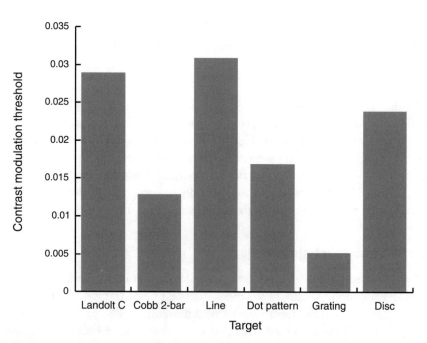

Figure 6.8 Comparison of threshold modulation required for detail to be judged visible. (Data from [1].)

lines are analog signals. A variety of other formats also exist. For example, PAL, a 625-line system at 50 Hz, is common in Europe [16].

Because the video image is sampled in one dimension (vertical), the effect of this sampling on resolution needs to be considered. A resolution target aligned with the scan lines may show three different resolutions, depending on how the target bars are aligned with the scan lines (Figure 6.9). A value called the Kel factor has been used to account for this effect [17]. The Kel factor indicates that the effective resolution of a sampled system is 0.7 times that of a continuous system. A further complication arises in that, whereas resolution line pairs are used for photographic imagery, TV lines are used for video.

Applying the Kel factor, National Television Systems Committee (NTSC) video has 343 effective lines in the vertical dimension. NTSC video uses a format where the vertical dimension is 3/4 of the horizontal (4:3 aspect ratio format). If horizontal resolution is assumed equal to vertical, NTSC video has 457 effective horizontal elements. The actual number of resolvable elements is a function of the system's frequency response. Much of the early work done with S&R video used lines on target as the measure of resolution. This was generally defined in terms of active lines, but did not account for sampling effects (see, for example, [18]).

One of the results of the work on video was the development of rules of thumb for required resolution. These rules were expressed in terms of number of subtending lines, generally across the minimum dimension of the target. Probably the most often quoted is the work by Johnson [19]. Johnson was actually concerned with image intensifiers and performed a study in

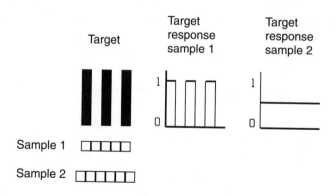

Figure 6.9 Effect of sampling on resolution target reading.

Table 6.2

Johnson Criteria

Target broadside view	Resolution per minimum dimension			
	Detection	Orientation	Recognition	Identification
Truck	0.90	1.25	4.5	8.0
M-48 tank	0.75	1.2	3.5	7.0
Stalin tank	0.75	1.2	3.3	6.0
Centurion tank	0.75	1.2	3.5	6.0
Half-track	1.0	1.5	4.0	5.0
Jeep	1.2	1.5	4.0	5.0
Command car	1.2	1.5	4.3	5.5
Soldier (standing)	1.5	1.8	3.8	8.0
105 Howitzer	1.0	1.5	4.8	6.0
Average	1.0 ± 0.25	1.4 ± 0.35	4.0 ± 0.8	6.4 ± 1.5

which observers were required to detect, determine the orientation of, recognize, and identify a variety of military equipment. Details of the experiment were not reported with the results. Table 6.2 shows results for detection, recognition, and identification. Johnson reported results in terms of equivalent resolving power (i.e., line pairs), but his results are sometimes misquoted as subtending lines. Johnson's data also assumes that contrast and signal-to-noise ratio (SNR) are adequate.

The use of subtending lines of resolution is generally adequate with NTSC video, but is somewhat less so with evolving digital formats such as ITU-RBT.601-4 [20]. Such systems will be presumably treated in the same manner as sampled EO systems.

6.1.4 Electro-optical Systems

Electro-optical systems include visible and IR sensors collecting imagery sampled in at least one dimension. Early EO systems were IR line scanners. They were sampled in the along-track dimension (Figure 6.10) and were continuous analog in the cross-track. With these early systems, resolution was defined in terms of the instantaneous field of view (IFOV), as shown in Figure 6.11. This was an angular measure (usually in milliradians) relating to the size of the detector and the optics. A 1-milliradian system at 1000-ft altitude has a detector footprint of 1 ft directly beneath the sensor. As the angle

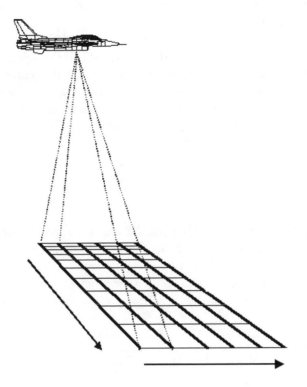

Figure 6.10 IR line scanner.

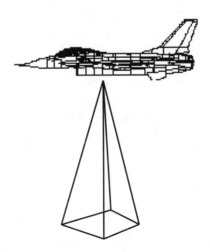

Figure 6.11 IFOV definition.

between the pointing angle and vertical increases, the size of the projected detector on the ground increases.

With systems sampled in two dimensions, the resolution measure called ground sampled distance (GSD) was developed. GSD is the detector or pixel pitch dimension projected to the ground. Pixel pitch is the center-to-center distance between adjacent picture elements (pixels). GSD indicates the energy footprint that will be sampled and is only a rough indicator of the size of objects that can be resolved. GSD is defined in one dimension as

$$\text{GSD} = \frac{\left[(p/f) \times SR\right]}{\cos(LA)} \tag{6.7}$$

where p is pixel pitch, f is focal length, SR is slant range (all three values measured in the same units), and LA is the angle (look angle) between the ground plane and the line of sight between the target and the sensor measured in degrees. In two dimensions, GSD is the geometric mean of the along-scan and cross-scan GSD, corrected if necessary where the two scan directions are not orthogonal:

$$\text{GSD}_{GM} = \sqrt{(\text{GSD}_X \times \text{GSD}_Y \times \sin\alpha)} \tag{6.8}$$

where α is the angle between the along-scan and cross-scan direction.

Figure 6.12 defines look angle. At some as yet undefined look angle, GSD must be measured in the plane orthogonal to the line of sight rather than in the ground plane. For ground-based and very long range tactical systems, measurement of GSD in the ground plane leads to meaningless values in one dimension. It is also the case, however, that for a large number of objects, the plan view conveys more information than an elevation. There is a target-type dependency and also a target-orientation dependency.

Acutance as a measure of edge sharpness has also been applied when EO imagery is written to paper or film. The EO digital display analog to acutance is the relative edge response (RER). RER is a measure of the slope of a normalized edge response (Figure 6.13). It is defined by the slope of the response ±0.5 pixel from the edge for a noise-free signal. The normalized edge response (ER) is calculated by

$$ERx(d) = 0.5 + \frac{1}{\pi}\int_{o}^{(nOptcutx)}\left[\frac{SystemX(\xi)}{\xi} \times \sin(2\pi\xi d)\right]d\xi \quad \text{X-axis} \tag{6.9}$$

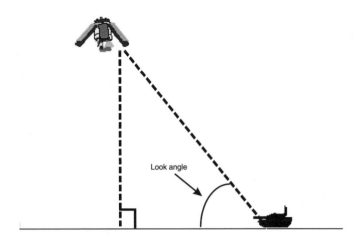

Figure 6.12 Look angle.

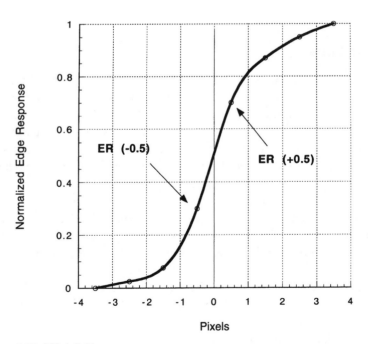

Figure 6.13 RER definition.

$$ERy(d) = 0.5 + \frac{1}{\pi} \int_{o}^{(nOptcuty)} \left[\frac{SystemY(\xi)}{\xi} \times \sin(2\pi\xi d) \right] d\xi \quad \text{Y-axis} \quad (6.10)$$

where $nOptcutx$, $nOptcuty$ = normalized (to effective sample spacing) optics cutoffs, $SystemX$, $SystemY$ = system MTF in X and Y direction, d = response position from the center of a horizontal pixel, and ξ = spatial frequency in cycles per sample spacing.

RER is defined as the difference between points -0.5 and $+0.5$ pixels from the center of the edge. Thus,

$$RER_x = ER_x(0.5) - ER_x(-0.5)$$
$$RER_y = ER_y(0.5) - ER_y(-0.5) \quad (6.11)$$

The geometric mean RER is calculated by

$$RER_{GM} = \sqrt{(RER_x \times RER_y)} \quad (6.12)$$

6.1.5 SAR Resolution

The primary measure of SAR resolution is the -3 dB (half power) impulse response (IPR) width measured in the ground plane [21]. This is shown in Figure 6.14. Two point returns separated by less than the IPR width cannot be resolved. Because of destructive and constructive wavefront interference, objects separated by greater than the IPR width may not always be separated. It is perhaps for this reason that the use of reflector resolution arrays has fallen out of favor. Such arrays consist of generally corner reflectors (Figure 6.15) arrayed in a geometric pattern.

A single reflector of known radar cross-section (RCS) can be used to measure IPR. Radar reflectors bounce transmitted energy back to the radar receiver, where it ultimately is used to create an image. The goal in measurement is to image only the energy reflected from the target and not that reflected from other objects in the scene. Similarly, the target must be sufficiently large in cross-section such that noise does not contribute significantly to the measurement. Trihedral corner reflectors (Figure 6.16) are an efficient means of achieving this goal.

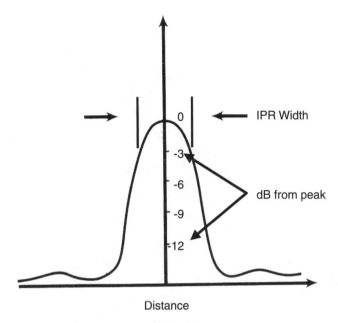

Figure 6.14 Impulse response.

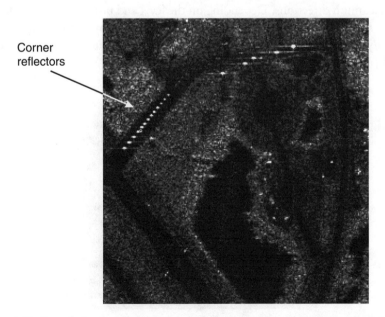

Figure 6.15 Corner reflector array. (©2000 Veridian Systems/Veridian ERIM International, Inc.)

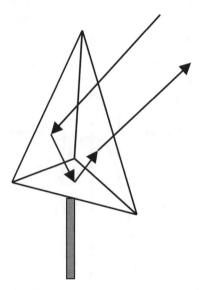

Figure 6.16   Trihedral corner reflector.

Although not a measure of resolution, sidelobe levels affect the appearance of point returns (Figure 6.17). Sidelobes can occur in both the range and azimuth directions. Depending on the level and positioning of sidelobes, they may interfere with other point returns or they may enhance the detectability of isolated point returns. They can also soften the appearance of point returns and affect detectability.

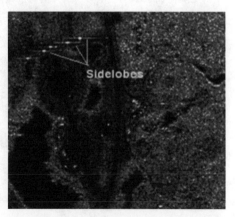

Figure 6.17   SAR sidelobes. (©2000 Veridian Systems/Veridian ERIM International, Inc.)

6.2 Contrast and Noise

Contrast and noise measures relate to the ability to detect tonal differences between objects of interest and backgrounds. Contrast is a measure of an average tonal or energy difference. The effect of contrast on resolution has previously been discussed. The term "contrast modulation" was introduced. Although it is perhaps the most common measure currently used, other measures have also been used. Contrast measures can be applied globally to a scene, or they can be applied to specific objects within a scene.

An image of a homogeneous area in a scene is not itself homogeneous. The image is subject to random energy fluctuations and is thus said to appear noisy. The cause of these random fluctuations varies with sensor type. Noise interferes with the ability to detect detail in a scene. Figure 6.18 diagrams the effect. As the random fluctuations increase, the ability to detect the same mean signal difference decreases. Noise may be measured itself, or it can be measured in relation to a defined signal.

6.2.1 Photographic Measures

In its most simple form, contrast is a ratio typically defined in luminance space. Contrast is a function of reflectance (the percentage of energy reflected from a surface), incident sunlight illumination, incident skylight illumination, atmospheric transmission, and haze. Skylight is energy incident on the target resulting from sunlight scattered in the atmosphere. Haze is scattered sunlight incident on the lens of the camera. Figure 6.19 diagrams the relationships. The luminance (L) reaching the camera can be defined as [22]

$$L = (S_u + S_s)(T)(R) + H \qquad (6.13)$$

where S_u = sunlight illuminance, S_s = skylight illuminance, T = atmospheric transmission (%), R = object reflectance (%), and H = haze luminance. Illuminance (incident energy) is measured in lux (lumens per square meter) and luminance (energy per unit of projected area per solid angle arriving at or leaving a surface in a defined direction) in candelas per unit square area (cd/m^2). Contrast (C) is defined as the ratio of two luminance values:

$$C = L_1/L_2 \qquad (6.14)$$

and is thus a function of the reflectance properties of two surfaces, because incident illuminance is the same for all surfaces in a scene at the same angle

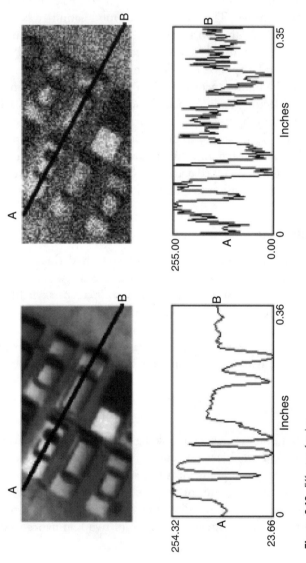

Figure 6.18 Effects of noise.

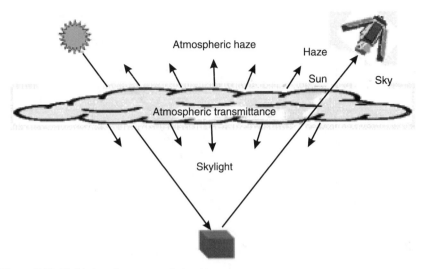

Figure 6.19 Visible imaging energy relationship.

relative to incident sunlight. Note that haze adds equally to the luminance from both surfaces and thus has the effect of reducing contrast.

Contrast is also a function of the photographic transfer process. This transfer process defines the relationship between energy incident on the camera lens or film and film transmittance (the percentage of light that will pass through the film). The camera lens both reduces the energy reaching the film and adds a haze component due to reflections within the optical train. The process is defined in terms of the relationship between log exposure and log transmittance. Log transmittance is inversely related to film density [1]:

$$D = \log_{10}\left(\frac{1}{T}\right) \tag{6.15}$$

where D = density and T = transmittance.

Figure 6.20 shows a typical function. The function is called an H (exposure) and D (density) curve after Hunter and Driffield [23]. The slope of the curve is called gamma. Gamma is defined as

$$\gamma = (\Delta D / \Delta \log H). \tag{6.16}$$

Gamma is thus a relative measure of contrast. For a given range of exposure, a higher gamma will show a greater range in transmittance and thus contrast.

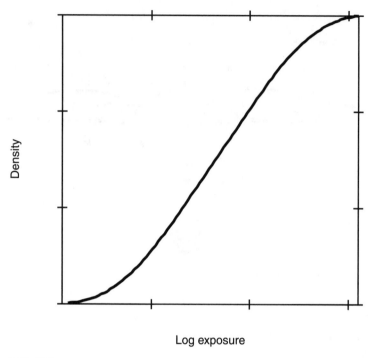

Log exposure

Figure 6.20 H&D curve.

Studies on the effects of contrast have shown mixed results, partly be-cause of the difficulty of defining meaningful measures of contrast. Objects and backgrounds in a scene are not homogeneous, and thus it is difficult to define a valid measure relating target/background contrast. Where contrast was defined as the average density difference between a target and its back-ground, no significant contrast effect was found [24]. When maximum density was reduced by pre-exposure fogging, a contrast effect on target com-pleteness was found only for small-scale images [25]. Where contrast was defined as the average of the six largest monotonic density variations across a scene, a weak correlation with identification accuracy of 0.32 was found [26]. Two additional studies [27,28] also found no relationship between tar-get and background brightness differences. Greater success in showing the effect of contrast was shown when photographic gamma was varied over the range 1.02–3.95 [29]. The images were projected under different levels of veiling light (simulating haze effects) to produce a display gamma range of 0.1–4.0. Results showed a performance asymptote at a display gamma of ~1.5. With photographic systems, film grain is the dominant noise source. The random distribution of silver grains in the photographic emulsion produces random variations in film density. The physical measure is granularity; the perceptual phenomenon is graininess. Granularity is defined as [30]

$$G = 1000 * \sqrt{\Sigma(D_m - D)^2/N} \qquad (6.17)$$

where D_m is average density, D is the density of each individual measurement, and N is the number of measurements. Measurements are made with a densitometer having an f 2.0 system aperture and a 48-μm diameter circular scanning aperture. The value of granularity reportedly correlates with the subjective appearance of graininess when viewed at 12× magnification.

In a study of target identification [26] no correlation with performance was found over granularities ranging from 20 to 160. When simulated film grain size exceeded 25–30% of a ground resolution element, performance was affected [24]. Film grain is thus generally an issue only for high-resolution small-scale systems. It may also be an issue with very high speed films designed for low light level applications, as well as multispectral systems where only a portion of the incident energy is used to expose each wavelength band.

6.2.2 Electro-optical Measures

Early EO systems recorded on film, and thus, photographic contrast and noise measures were applicable. For digital systems, the tendency has been to use SNR to express sensitivity and MTF to express throughput contrast. Signal is the output of a detector measured either at the detector or at the display. Noise is random signal variations due to detector noise, photon or shot noise, temperature noise, and amplifier noise. Two measures are commonly used. The first is signal-to-noise ratio defined as

$$S/N = \frac{S}{\left(\Sigma_{i=1}^{n} (S_{avg} - S_i)^2 / n \right)^{1/2}} \qquad (6.18)$$

where S is the output of the detector for a given level of incident energy, S_{avg} is the average signal, S_i is the instantaneous signal level, and n is the number of measurements or samples [2]. As S/N decreases, detectability decreases. A study of flight navigation checkpoint recognition on a TV display [31] showed performance degraded below a S/N of 1.2. A study using artificial targets showed a decrease in probability of recognition at a S/N of 3.5 [32]. In the examples given, S/N was expressed as a ratio. S/N is also commonly expressed as a value in decibels. S/N in decibels (dB) is defined as

$$S/N_{dB} = 20 \log_{10} S/N. \tag{6.19}$$

Two measures related to S/N are commonly used to describe the sensitivity of detectors. They are spectral noise equivalent power (NEP), or irradiance, and noise equivalent delta temperature, or more generally, noise equivalent irradiance. Spectral noise equivalent power is defined as [2]

$$NEP(\lambda) = N/R(\lambda) \tag{6.20}$$

where N = noise and $R(\lambda)$ = spectral responsivity. It is the incident flux equivalent needed to provide an SNR of 1. Noise equivalent delta temperature (NEΔT) is defined as the temperature change of a blackbody necessary to achieve a S/N of 1.

Although less commonly used, bit depth and dynamic range are two contrast-related measures applicable to EO imagery. Dynamic range is defined as the energy range over which the system response is monotonically changing. At some point, the energy level may exceed the capability of the detector to provide an increased response. At that point, saturation occurs, resulting in an inability to separate different levels of energy even when they exceed the system S/N threshold. Dynamic range is defined as the ratio of the maximum signal to system noise (a S/N ratio) and is usually expressed in decibels.

"Bit depth" is a term applicable to digital systems and refers to the number of levels used to record and display signals. An 8-bit system has 256 (2^8) discrete levels; an 11-bit system has 2048 (2^{11}) levels. For a given dynamic range, as the bit depth is decreased, greater quantization error (noise) occurs. Regardless of the noise performance of the detector, the display is limited to the number of levels defined by bit depth.

6.2.3   SAR Measures

SAR imagery, because it is coherent, shows what is called clutter. Clutter is represented by random returns from diffuse features. Features such as vegetation, soil, and pavement have (relative to the SAR wavelength) a rough surface containing many small reflectors. These reflectors produce wavelets with random phase. These wavelets interfere with each other, both constructively and destructively. The effect is a random pattern of speckle, which represents power from a distributed target. Clutter takes on different mean intensities, depending on the material and the angle of energy incidence. Figure 6.21 shows an example of SAR imagery. Clutter returns range from

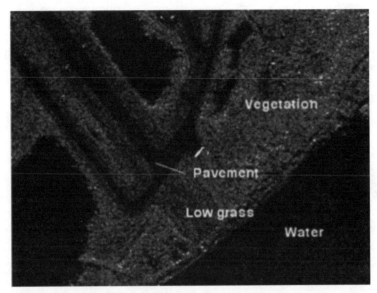

Figure 6.21   SAR image exhibiting different clutter levels. (©2000 Veridian Systems/Veridian ERIM International, Inc.)

very bright in areas of vegetation to negligible in water areas. Table 6.3 shows measured clutter returns as a function of aspect angle and material. Clutter variance is proportional to mean clutter intensity.

If trying to find a target return in an area of clutter, the clutter acts as noise. The different intensities of clutter also provide information (the RCS of the material can be estimated), and in that sense, clutter is a signal. The variation in the clutter signal is a form of noise called multiplicative noise. Because clutter variance is proportional to the mean, the clutter noise increases with the mean and is thus multiplicative. Additive noise also occurs

Table 6.3
Background Feature Cross-Section Values (X-band, dB)

Material	Grazing angle (degrees)				
	10	**20**	**30**	**40**	**50**
Trees	−13.5	−13.5	−13.1	−12.8	−11.6
Shrubs	−15	−13	−1	−10.5	−9
Short vegetation	−15.5	−13.5	−12	−11	−9.5
Grass	−15.9	−14.1	−13.	−12.3	−11.4
Concrete	−47	−36	−32	−29	−27

with SAR. The RCS measurement of clutter is biased by the effect of additive noise. A variety of noise-related measures can be formulated [33].[1] We begin with the clutter-to-noise ratio (CNR) measure as an indication of the biasing effect of additive noise:

$$\text{CNR} = \frac{|C|^2 \, \sigma^o R_{h_a}(0)}{NR_l(0)} \tag{6.21}$$

where C is range-dependent gain, σ^o is the clutter RCS, R_{h_a} is the azimuth point spread function peak energy (assuming that the peak occurs at $x = 0$) at range R, N is the noise power, and $R_l(0)$ is the energy related to the operation of the SAR processing filter. The signal-to-clutter ratio (SCR) is expressed as

$$\text{SCR} = \frac{\sigma \, |h_a(0)|^2}{\sigma^o R_{h_a}(0)} \tag{6.22}$$

where σ is the target RCS, $h_a(0)$ is the peak response from the target of RCS σ, and clutter is as in (6.21).

Finally, the signal-to-noise ratio is defined as

$$\text{SNR} = \text{CNR} \times \text{SCR} = \frac{|C|^2 \, \sigma \, |h_a(0)|^2}{NR_l(0)} \tag{6.23}$$

where all terms are as previously defined. SNR can also be defined as in (2.15).

Because the mean power in a processed signal is of the form clutter plus noise, a target-to-background ratio (TBR) can also be formed:

$$\text{TBR} = \frac{|C|^2 \sigma \, |h_a(0)|^2}{|C|^2 \sigma^o R_{h_a}(0) + NR_l(0)} \tag{6.24}$$

---

1. Equations (6.21)–(6.23) reprinted from *Understanding Synthetic Radar Images* by C. Oliver and S. Quegan, Artech House Inc., Norwood, MA, www.artechhouse.com.

The different SAR measures have different applications. If the task is one of mapping terrain, then CNR is most appropriate. Where clutter is low and target detection is of interest, SNR is most appropriate. Where clutter is high and noise is low, SCR may be appropriate, or, where both clutter and noise are of interest, TBR.

The dynamic range and bit depth metrics defined for EO systems are also applicable to SAR. They are defined similarly, although dynamic range is typically defined in RCS space. RCS is the size of a flat plate reflector that would provide a return intensity equal to an observed return. An RCS of $1m^2$ means that the return is equivalent to that that would be received from a specular reflector $1m^2$ orthogonal to the radar beam. A trihedral reflector, for example, will have an RCS much larger than its physical dimensions.

6.3 Artifacts

An artifact is defined as a structure or substance not normally present. Image artifacts can occur from a wide variety of causes. They include sampling, detector gain control, bandwidth compression, and image mosaicing. With a few exceptions, however, artifacts have not been quantified in terms of their effect on information extraction performance. Similarly, there are few measures to quantify the physical nature of the artifacts. The major exception is in the field of video data compression. Here, considerable work has been done on quantifying each of the potential compression artifacts. Although not performed in the context of S&R systems, this work has potential application, particularly with digital videos systems, and is briefly reviewed.

With digital systems, undersampling produces aliasing. Aliasing, in a line scan system, is the generation of a spurious spectrum resulting from sampling at rates below two samples per cycle, the Nyquist criteria [34]. The spurious spectrum, if proper filtering is not applied, results in false information in the scene. This false information may appear as banding (Moiré patterns), blurring, jagged edges, or enhanced edges. The effect of this spurious information is highly scene dependent and also task dependent. It is perhaps for this reason that there is no universally accepted measure of aliasing. It is recognized that undersampling degrades interpretability [35,36]. A study of target recognition [36] quantified spurious responses in terms of in-band (edge shifting, line width variations) and out-of-band (raster, sharp pixel edges). An MTF correction function (MTF squeeze) was related to target recognition and identification performance. Targets were large-scale elevation views (IR images) of armored vehicles. It was found that out-of-band spurious responses (horizontal LOS) were more important for identification than in-band; both were important for recognition. Aliasing has also been treated as a noise effect [37]. Aliasing is not always evident as obviously spurious responses; it may show up as high-frequency energy characterized as scene-dependent noise.

Schott [2] discusses artifacts in terms of detector calibration errors or failures in linear arrays and whiskbroom detectors. The result is banding on the image. Although the effect is objectionable and can be quantified in terms of noise, the effect on interpretability has not been defined.

A similar situation exists with mosaicing errors. EO staring systems acquire many individual images that are mosaiced together to form a single large image. Two types of errors may occur. The first is spatial—the sub-images may be displaced in x or y, resulting in image gaps, overlap, or lateral displacement. Although each of these errors can be quantified, the impact on interpretability depends on two interrelated factors. The first is the relationship between the size of objects of interest and the mosaicing error. The second is the location of objects of interest relative to the errors. If objects are small and fall on or in the region of error, interpretability is degraded. If they do not fall in the region of error, the error has little or no effect. Large objects (relative to the error) are generally not affected.

The second type of mosaicing error occurs because of a gain or contrast mismatch. The result is a contrast change across the mosaic boundary. Although the error can be quantified in terms of an SNR or contrast change, the effect on interpretability is seldom significant unless the error is sufficient to substantially degrade SNR. It is thus the gain error rather than the mosaic error that is important.

Bandwidth compression can introduce a variety of spatial and tonal artifacts. In the case of video, temporal artifacts can also be produced. It is in the video arena that much of the work on metric development has taken place. Table 6.4 characterizes common artifacts resulting from video com-

Table 6.4
Video Compression Artifacts

Type	Example	Definition
Resolution degradation	Blocking	Rectangular or checkerboard patterns
	Blurring/smearing	Loss of detail and edges
	Jerkiness	Loss of smooth motion
Edge busyness	Mosquito noise	False movement of small moving objects
Image persistence	Erasure	Erased object continues to appear

Horizontal　　　　　　Vertical

-1	-2	-1
0	0	0
1	2	1

-1	0	1
-2	0	2
-1	0	1

Figure 6.22　Sobel filters used by Wolf [38].

pression [38]. These artifacts occur in the luminance channel; additional chrominance artifacts also occur.

One common artifact metric has its origin in edge detection. A filter of the type shown in Figure 6.22 [38] is applied to detect edges. A measure based on edge differences between original and compressed images is then defined and related to DSIS or DSCQS ratings. It can also be used to quantify blurring and blocking. This same approach should be applicable to static imagery.

A variety of other measures have been proposed and evaluated. Pessosa et al. [39], for example, segments images into edge, plane, and texture areas and computes mean-square-error and Sobel differences to characterize bandwidth compression effects relative to subjective estimates. A mean absolute error of less than 4% was computed for five scenes processed through 26 bandwidth compression algorithms. The error was computed relative to a normalized impairment scale difference. Webster et al. [40] developed a spatial metric based on the Sobel filter defined as

$$SI[F_n] = STD_{Space}\{Sobel\,[F_n]\} \qquad (6.25)$$

where F_n is the Sobel filtered video frame at time n, and STD_{Space} is the standard deviation over the horizontal and vertical dimensions of the frame. A temporal measure was defined as

$$TI[F_n] = STD_{Space}[\Delta F_n] \qquad (6.26)$$

where ΔF_n is the difference between pixel values at the same location but at successive times of frames, and STD_{Space} is the standard deviation over the horizontal and vertical dimensions of the frame. ANSI T1.801.03-1996 [41] describes a variety of related spatial and temporal measures as well as the measurement procedures.

6.4 Summary Measures

Up to this point, generally unitary measures of resolution, contrast, and noise have been discussed. All have been shown, at least over some range, to affect information extraction or interpretability performance. A variety of measures have been developed in an attempt to combine these three unitary measures into a single summary measure. These measures often include some expression to account for characteristics of the human visual system. Most of these measures are based on the concept of MTF or are in some way related to MTF.

6.4.1 Regression Models

Despite several studies in which multiple image quality variables were studied [24–27], the development of multiple regression models combining different metrics was not common. One exception is noteworthy [24]: a multiple regression equation combining measures of contrast, granularity, and resolution. An R^2 value of 0.86 was reported for a four-term equation with linear terms for contrast and resolution and a linear and quadratic term for granularity. Unfortunately, the variations in quality were achieved using photographic degradation, and thus, the model was of no value as a design tool. It was useful, however, in indicating the relative importance of the physical variables.

6.4.2 MTF

In Chapter 2, the point spread function and its Fourier transform, the MTF were introduced. The modulation transfer function defines contrast modulation (Cm) as a function of spatial frequency [2,43]. As spatial frequency increases, the ability of a system to transfer or portray contrast differences diminishes. MTF is a useful measure in that it relates contrast and resolution. It also forms the basis for defining the edge response. As a predictor of performance, it has generally been combined with other measures.

MTF is classically measured with a sine wave target (Figure 6.23). As the sine wave frequency increases, contrast (and thus contrast modulation)

Figure 6.23 Sine wave target.

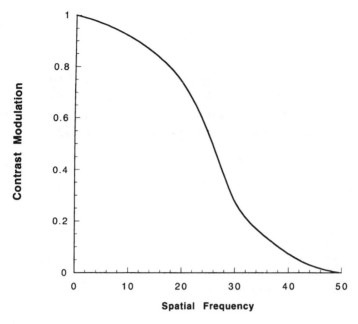

Figure 6.24 MTF.

decreases. A plot of Cm versus spatial frequency (Figure 6.24) defines the MTF of a transfer process. The transfer process is a transfer of energy through a medium or display.

As noted previously, the MTF can also be derived from the system edge or point spread function. If the two dimensions of the system (x and y) are separable, the MTF is the Fourier transform of the line or point spread function or impulse response [2]. The MTF of a system is a function of several elements in the imaging chain and can be derived from successive multiplications of the contributing MTFs. A Cm value of 0.8 and a Cm value of 0.6 (for a given spatial frequency and two system components) would yield a Cm of 0.48. This is true only for an incoherent system (such as an EO system), where phase effects can be ignored.

Sources of contrast loss include the atmosphere (generally only at high frequencies), lens systems, detectors, and recording or display devices. As will be shown in the next section, the MTF of the human visual system can also be considered. The system MTF is produced by cascading (multiplying) the individual MTFs. Figure 6.25 shows a hypothetical example.

A system MTF has a cutoff frequency where Cm = 0. For an optical system, the cutoff frequency is defined as

$$\xi_{Max} = \frac{d}{\lambda f} = \frac{1}{FN\lambda} \qquad (6.27)$$

where d = diameter, λ = wavelength, f = focal length, and FN = f-number.

Closely related to the MTF is the contrast transfer function (CTF). The CTF also plots modulation as a function of spatial frequency but is defined using square wave targets. The CTF and MTF are related [2]:

$$CTF_f = \frac{4}{\pi}\left[MTF(f) - \frac{MTF(3f)}{3} + \frac{MTF(5f)}{5} + \frac{MTF(7f)}{7} + \cdots \right]$$

$$MTF_f = \frac{\pi}{4}\left[CTF(f) - \frac{CTF(3f)}{3} + \frac{CTF(5f)}{5} + \frac{CTF(7f)}{7} + \cdots \right] \qquad (6.28)$$

where f is spatial frequency in cycles per millimeter.

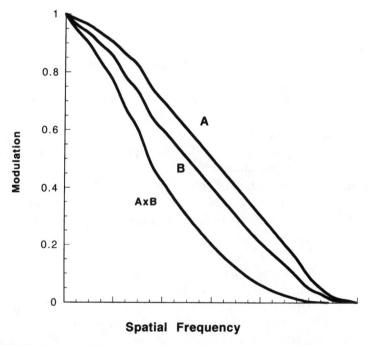

Figure 6.25 Cascaded MTFs.

6.4.3 MTF Measures

A variety of summary measures have been developed based on the MTF. One of the first was the threshold quality factor (TQF) developed by Charmin and Olin [44]. Borough et al. renamed it the modulation transfer function area (MTFA) [3]. The area between the MTF and detection threshold function (Figure 6.26) defines the MTFA for a photographic system. MTFA was defined as

$$\text{MTFA} = \int_0^{k_1} \left[T(k) - \frac{M_t(k)}{M_0} \right] dk \qquad (6.29)$$

where k_1 is the spatial frequency where the detection threshold and MTF curves intersect, M_0 is the object contrast modulation, $T(k)$ is the MTF at frequency k, and $M_t k$ is the detection threshold of the eye at frequency k.

The detection threshold function is defined by a combination of the modulation threshold of the human visual system (based on visual system noise) and film grain. For a film system, the detection threshold curve was defined as [45]

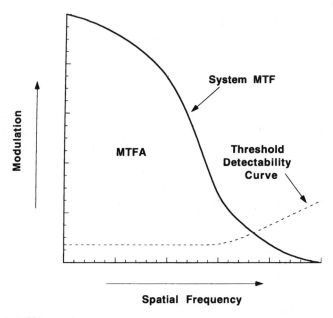

Figure 6.26 MTFA.

$$M_t(k) = 0.034\gamma^{-1}[0.033 + \sigma(d)^2 k^2 s^2]^{0.5} \qquad (6.30)$$

where 0.033 is an empirically derived constant, γ is the film gamma, 0.034 is an empirically derived constant, $\sigma(d)^2$ is film granularity, k is spatial frequency, and s is the S/N ratio for threshold viewing (assumed to be 4.5).

The MTFA was shown to correlate well with photointerpreter paired comparison rankings of image quality. The correlation was 0.92 for the linear form and 0.90 for the log form. A subsequent study [46] using a nine-point scale of interpretability (early version of NIIRS) showed a correlation of 0.97 when data was averaged across scenes.

When attempts were made to apply MTFA to line scan systems (TV displays), some difficulties were encountered in that one dimension (along scan) was continuous and the other was sampled. MTF was measured in two dimensions for a TV display, and two measures combined in various ways to relate to performance in a target recognition task [47]. Correlations were lower than reported for photographic imagery, possibly because of the measure, but also because target search was required. The two-dimensional form, however, outperformed the one-dimensional form.

Several measures related to MTFA but designed to overcome perceived limitations of the MTFA concept have been evaluated [48]. First was the integrated contrast sensitivity function [49]. This was based on the recommendation that the MTF be multiplied by the contrast sensitivity function (CSF) of the visual system and was defined as

$$ICS = \int_0^\infty \frac{MTF_\omega}{CSF_\omega}\,d\omega \qquad (6.31)$$

where MTF is the display MTF at frequency ω (in angular units), and CSF is the contrast sensitivity at frequency ω. The CSF defines modulation thresholds as a function of spatial frequency. A measure called the subjective quality factor (SQF) was developed based on the observation that the visual response is mostly limited to the spatial frequency range of 3 to 12 cycles per degree [50]. The SQF was defined as

$$SQF = K \int_{\ln a}^\infty \int_0^{2\pi} \left| M(\ln \omega, \theta) \right| \left| M_v(\ln \omega) \right| d\theta d(\ln, \omega) \qquad (6.32)$$

where $M(\ln,\omega,\theta)$ is the two-dimensional optical transfer function of the display system (polar coordinates), $M_v(\ln,\omega)$ is the MTF of the visual system, ω is angular spatial frequency, and θ is orientation angle of spatial frequency.

To assess these metrics, a series of recognition studies was performed, one on a TV display and one on a film display [48]. Figure 6.27 shows results for the two recognition studies. Results are shown as the correlation of the value of the physical metrics with recognition performance for a TV and a film target recognition study. In addition to MTFA (log of linear MTFA), SQF, and ICS, results for a band-limited MTFA (MTFA/BL) are shown. The upper limit of the band-limited MTFA was defined by a sine-wave and square-wave discrimination function developed by Campbell and Robson [51]. It denotes the ability to distinguish between a sine wave and a square wave. The lower limit was set at 1 cycle per degree. For the TV study, the band-limited MTFA showed significantly higher correlations than the other measures; for the film study, differences were not statistically significant.

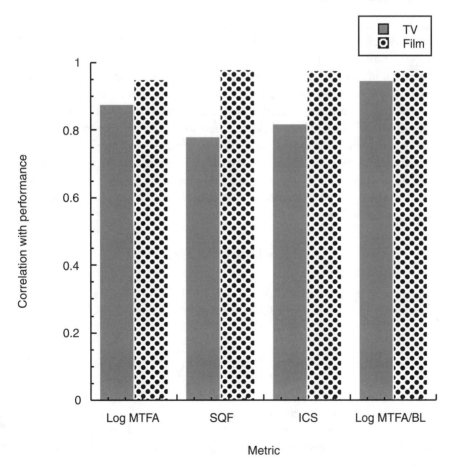

Figure 6.27 MTF metrics versus performance. (Data from [48].)

Somewhat related to the MTFA concept is the JND model developed by Carlson and Cohen [52]. This model defined a just-noticeable difference in terms of display MTF differences. A metric based on the JND model was developed by computing the area under the JND curves. It was called the JND area (JNDA). When the lower integration limit was set at 1/2 cycle per degree, the JNDA performed as well as the band-limited MTFA metric for the two recognition studies. The Carlson and Cohen work forms the basis for the Sarnoff JND model discussed in Chapter 10.

Directly related to the MTFA and the TQF is the radar threshold quality factor (RTQF) developed by Mitchel [53]. This model will be discussed in more detail in Chapter 10, but it is simply an extension of the TQF to the radar domain, while accounting for radar noise and clutter.

6.4.4 SNR Measures

SNR and related metrics were previously discussed as physical quality measures. Rosell and Willson [54] developed a metric called the display signal-to-noise ratio (SNR_D). The measure is based on a large body of work on signal detection in noise. It assumes that the detection process of a transducer (detector) is subject to statistical fluctuation. The theory of signal detection discussed in Section 5.3 is used to explain behavior. It follows that some threshold SNR (related to d′) is required for detection to occur. Much of the original work was based on detection of simple geometric targets, but Rosell and Willson extended the approach to more meaningful targets using the detection and recognition data developed by Johnson. The development of the concept is documented in Biberman [54]. In its simplest form, SNR_D is defined as

$$SNR_D = [2t\Delta f(a/A)]^{1/2} \times SNR_V \qquad (6.33)$$

where t = integration time, $2\Delta f$ = video bandwidth in Hertz, $2\Delta f$ = image area/photo surface area, and SNR_V = video SNR. Video SNR is defined as

$$SNR_V = \frac{2MG_T G_V i_{av}}{(2ei_{av}G_T^2 G_V^2 \Delta fv + 2eG_V^2 i_s \Delta fv)^{1/2}} \qquad (6.34)$$

where $2M$ = image contrast, G_T = sensor target gain, G_V = video gain, i_{av} = average photocurrent, e is the charge of an electron (coulombs), and i_s = average added noise photocurrent. This equation is valid if the sensor spatial frequency response is not a limiting factor and also assumes a perfect display. Adjustments are made to account for finite apertures, finite noise spectrum, and target shape effects.

After showing the validity of the model for predicting detection probabilities for simple geometric shapes, Johnson's data [19] was used to extend the model to real-world targets. If the SNR_D model indicated that a two-line pattern could be detected, following Johnson's criteria, it should be possible to detect a vehicle and, with 8 lines, to identify the vehicle. The bar pattern results are somewhat optimistic because of the degradation imposed by the sensor MTF. Figure 6.28 shows recognition results for three target types. Table 6.5 summarizes results for various recognition levels.

The SNR_D concept developed by Rosell and Willson is similar to the minimum resolvable contrast (MRC) and minimum resolvable temperature (MRT) metrics that are used in the Night Vision and Electronics Sensors Directorate (NVESD) model discussed in detail in Chapter 10. MRC is the contrast required to resolve a four-bar line pattern that is, in turn, related to

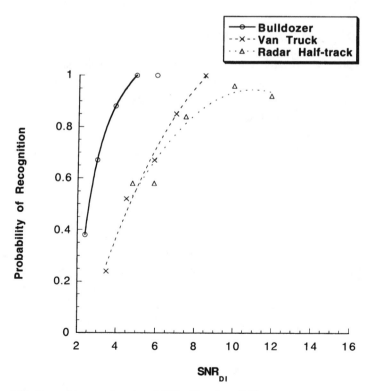

Figure 6.28 Recognition as a function of SNR_D. (Data from [54].)

Table 6.5
SNR_D Requirements for Various Levels of Discrimination

Level	Bkgd	Lines[1]	Threshold SNR_D for single bar of spatial frequency (lines/picture height) equal to		
			300	500	700
Detection	Uniform	1	2.8	2.8	2.8
Detection	Clutter	2	2.9	2.5	2.5
Recognition	Uniform	8	2.9	2.5	2.5
Recognition	Clutter	8	3.9	3.4	3.4
Identification	Uniform	13	3.6	3.0	3.0

[1]TV lines per minimum target dimension (height).

Johnson's recognition criteria. MRT is the same concept extended to IR systems.

6.4.5 Physical Difference Metrics

In Chapter 5, a difference metric called the JND metric was discussed [55]. The JND metric is based on a model of the human visual system and is used to predict direct and predicted information extraction performance. There are also several physical difference metrics commonly used to assess the effects of alternative processings of the same image. Probably the most common is the root mean square error (RMSE) metric. This is defined as

$$RMSE = \sqrt{\Sigma_{i=1}^{n}(I_O - I_D)^2/n \times m} \tag{6.35}$$

where I_O = original pixel value (amplitude), I_D = difference image pixel value, and n and m are the number of rows and columns of pixels, respectively. Although RMSE is easy to compute, it is generally not a good predictor of performance differences [56]. It is, however, sensitive to processing differences such as bandwidth compression. Closely related is the mean absolute error and peak signal-to-noise ratio (PSNR). PSNR is defined as

$$PSNR = 10 \log_{10}\left(\frac{CL^2}{MSE} \right) \tag{6.36}$$

where CL is the number of command or digital count levels, and MSE is the mean square error. Again, these measures are of limited value in predicting information extraction performance [57]. Other metrics include the Laplacian mean squared error and the Weighted Power Spectral Density metric [58]. In the previous discussion of video artifacts (Section 6.3), a variety of difference metrics were briefly discussed. Somewhat greater success in predicting performance has been shown using combinations of these difference metrics [39]. The spatial and temporal metrics defined in (6.18) and (6.19) were combined to form one spatial and two temporal metrics. The metrics were combined in a linear equation with coefficients empirically derived from application to a training set of 18 scenes and 36 video clips. When applied to an independent test set of equal size, the correlation with visual impairment ratings (DSIS) was 0.94. Similarly, three metrics (Laplacian mean-squared-error, average absolute error, and weighted power spectral density) were combined and showed good performance against predictions using the FLIR 92/Acquire model [58].

6.5 Summary

In this chapter, we have discussed a variety of physical quality measures or variables that have been used to predict some measure of direct or predicted information extraction performance. Simple metrics related to scale, resolution, contrast, and noise were all shown to affect performance or perceived quality. A variety of artifacts related to sampling, bandwidth compression, and mosaicing can also affect performance. Here, however, simple metrics are generally not available. In the video realm, some success has been achieved in the development of more complex metrics and models that quantify the effects of different types of artifacts. Several of these metrics appear applicable to static imagery but do not appear to have been used for this purpose.

Metrics that take into account the human visual system have also shown some success. The MTFA and related metrics combine physical characteristics of the image with the demands of the visual system. SNR-related metrics similarly combine image and visual performance characteristics.

Finally, a variety of metrics have been developed to quantify differences between two images resulting from system design or processing changes. Simple statistical comparisons (RMSE, PSNR) have generally not been accurate predictors of performance. More complex measures, particularly those that decompose the image into component parts, have shown greater success.

References

[1]   Farrell, R. J., and J. M. Booth, *Design Handbook for Imagery Interpretation Equipment,* Seattle, WA: Boeing Aerospace Co., 1984.

[2]   Schott, J. R., *Remote Sensing, The Image Chain Approach,* New York, NY: Oxford University Press, 1997.

[3]   Borough, H. C., et al, *Quantitative Determination of Image Quality,* D-2-114058-1, Seattle, WA: Boeing Aerospace Co., 1967.

[4]   Nill, N. B., and B. H. Bouzas, "Objective Image Quality Measure Derived from Digital Image Power Spectra," *Optical Engineering,* Vol. 31, No. 4, 1992, pp. 813–825.

[5]   Departments of the Army, the Navy, and the Air Force, *Photographic Interpretation Handbook,* Washington, DC: U.S. Government Printing Office, 1954.

[6]   Naval Reconnaissance and Technical Support Center, *Photointerpretation Handbook, Volume I,* TM 30-245, Washington, DC: U.S. Government Printing Office, 1967.

[7]   Riehl, K., and L. Maver, "A Comparison of Two Common Image Quality Measures," in *Airborne Reconnaissance XX,* SPIE, 1996.

[8]   *Reconnaissance Reference Manual,* Saint Louis: MO, McDonnell Aircraft Corporation, 1969.

[9]   Gliatti, E. L., "Image Evaluation Methods," in *Airborne Reconnaissance III,* SPIE, 1978, pp. 6–12.

[10] Kuperman, G. G., and E. L. Gliatti, "A Comparison of Two Subjective Image Quality Assessment Methods," *SPSE,* 1981.

[11] Maver, L. M., et al., "Aerial Imaging Systems," in J. V. Sturge, V. Walworth, and A. Shepp (eds.), *Imaging Processes and Materials,* Neblettes Eighth Edition, New York, NY: Van Nostrand Reinhold, 1989.

[12] Higgins, G. C., and R. N. Wolfe, "The Relation of Definition to Sharpness and Resolving Power in a Photographic System," *JOSA,* Vol. 45, No. 2, 1955, pp. 121–129.

[13] Brainard, R. W., and K. B. Caum, *Evaluation of an Image Quality Enhancement Technique,* AMRL-TR-65-143, Wright-Patterson Air Force Base, OH: Aerospace Medical Research Laboratories, 1965.

[14] Biberman, L. M., "Image Quality," in L. M. Biberman (ed.), *Perception of Displayed Information,* New York, NY: Plenum Press, 1973.

[15] ANSI/SMPTE 170-M, *Composite Analog Video Signal, NTSC for Studio Applications,* 1994.

[16] International Telecommunication Union, *Characteristics of Television Systems,* Report ITU-RBT.624-3, 1986.

[17] Jensen, N., *Optical and Photographic Reconnaissance Systems,* New York, NY: John Wiley and Sons, 1968.

[18]   Oatman, L. C., *Target Detection using Black and White Television, Study I: The Effects of Resolution Degradation on Target Detection*, Aberdeen Proving Ground, MD: U.S. Army Human Engineering Laboratories, 1965.

[19]   Johnson, J. "Analysis of Image Forming Systems," *Proceedings of Image Intensifier Symposium*, Ft. Belvoir, Virginia, 1959, pp. 269–273.

[20]   International Telecommunication Union, *Recommendation ITU-RBT.604-1, Encoding Parameters of Digital Television for Studios*, 1994.

[21]   Leachtenauer, J. C., R. H. Mitchel, and F. M. Damon, "Holographic Display of SAR Imagery," in *Airborne Reconnaissance III, Collection and Exploitation of Reconnaissance Imagery*, Vol. 137, pp. 172–178.

[22]   Maver, L., et al., "Aerial Imaging Systems," in J. Sturge, V. Walworth, and A. Shepp (eds.), *Imaging Processes and Materials*, New York NY, Van Nostrand Reinhold, 1989.

[23]   Hurter, F., and V. C. Driffield, "Photochemical Investigations and a New Method of Determination of the Sensitivty of Photographic Plates," *Journal, Society of Chemical Industrials*, Vol. 9, 1890, pp. 455–469.

[24]   Bennett, C. A., et al., *A Study of Image Qualities and Speeded Intrinsic Target Recognition*, Owego, NY: IBM Federal Systems Division, 1963.

[25]   Applied Psychology Corporation, *Performance of Photographic Interpreters as a Function of Time and Image Characteristics*, RADC-TR-63-313, Griffiss Air Force Base, NY: Rome Air Development Center, 1964.

[26]   Roetling, P. G., et al., *Quality Categorization of Aerial Reconnaissance Photography*, RADC-TDR-63-279, Griffiss Air Force Base, NY: Rome Air Development Center, 1963.

[27]   Brainard, R. W., et al., *Development and Evaluation of a Catalog Technique for Measuring Image Quality*, TRR 1150, Washington, DC: U.S. Army Personnel Research Office, 1966.

[28]   Jennings, L. B., et al., *Ground Resolution Study Final Report*, RADC-TR-63-224, Griffiss Air Force Base, NY: Rome Air Development Center, 1963.

[29]   Blackwell, H. R., J. G. Ohmart, and R. W. Brainard, *Experimental Evaluation of Optical Enhancement of Literal Visual Displays*, ASD TR 61-568, Wright-Patterson Air Force Base, OH: Aerospace Medical Laboratory, 1961.

[30]   Eastman Kodak Company, *Manual of Physical Properties, Aerial and Special Materials*, Rochester, NY: Eastman Kodak Company, 1967.

[31]   Radio Corporation of America, *Target Detection and Recognition Study*, Corona, CA: U.S. Naval Ordnance Laboratory, 1962.

[32]   Stathacapolous, A., et al., *Surveillance Systems Study*, Santa Barbara, CA: Defense Research Corp., 1967.

[33]   Oliver, C., and S. Quegan, *Understanding Synthetic Aperture Radar Images*, Norwood, MA: Artech House, 1998.

[34] Legault, R., "The Aliasing Problem in Two-Dimensional Sampled Imagery," in L. M. Biberman (ed.), *Perception of Displayed Information*, New York, NY: Plenum Press, 1973.

[35] Wittenstein, W., "Minimum Temperature Difference Perceived: A New Approach to Assess Undersampled Thermal Imagers," *Optical Engineering*, Vol. 38, No. 5, pp. 773–781.

[36] Vollmerhousen, R., R. G. Driggers, and B. O'Kane, "Influence of Sampling on Target Recognition and Identification," *Optical Engineering*, Vol. 38, No. 5, 1999.

[37] Park, S. K., and Z. Rahman, "Fidelity Analysis of Sampled Imaging Systems," *Optical Engineering*, Vol. 38, No. 5, 1999.

[38] Wolf, S., *Features for Automated Quality Assessment of Digitally Transmitted Video*, NTIA Report 90-264, Washington, DC: National Telecommunications and Information Administration, 1990.

[39] Pessosa, A., et al., "Video Quality Assessment Using Objective Parameters Based on Image Segmentation," *SMPTE Journal*, December 1969, pp. 865–872.

[40] Webster, A., et al., "Objective Video Quality Assessment System Based on Human Perception," in J. P. Allebach and B. E. Rogowitz (eds.), *Human Vision, Visual Processing and Digital Display IV*, Proc. SPIE, Vol. 1913, pp. 15–26.

[41] American National Standards Institute, *Digital Transport of One-Way Video Signals—Parameters for Objective Performance Assessment*, ANSI T1.801.03-1996, New York, NY: ANSI, 1996.

[42] Hollanda, P. A., and A. Scott, "The Informative Value of Sampled Images as a Function of the Number of Scans per Scene Object and the Signal-to-Noise Ratio," *Photographic Science and Engineering*, Vol. 14, No. 6, 1970, pp. 407–413.

[43] Driggers, R. G., P. Cox, and T. Edwards, *Introduction to Infrared and Electro-Optical Systems*, Boston MA: Artech House, 1999.

[44] Charmin, W. N., and A. Olin, "Tutorial-Image Quality Criteria for Aerial Camera Systems," *Photo. Sci. Eng.*, Vol. 9, No. 6, 1965, pp. 385–397.

[45] Snyder, H. L., "Image Quality and Observer Performance," in L. M. Biberman (ed.), *Perception of Displayed Information*, New York, NY: Plenum Press, 1973.

[46] Klingberg, C. L., C. S. Elworth, and C. R. Fileau, *Image Quality and Detection Performance of Military Photointerpreters*, D162-10323-1, Seattle, WA: Boeing Co., 1970.

[47] Snyder, H. L., *Visual Search and Image Quality*, AMRL-TR-76-89, Wright-Patterson Air Force Base, OH: Aerospace Medical Research Laboratory, 1976.

[48] Task, H. L., *An Evaluation and Comparison of Several Measures of Image Quality for Television Displays*, Wright-Patterson Air Force Base, Dayton, OH: Aerospace Medical Research Laboratory, 1979.

[49] van Meeteren, A., *Visual Aspects of Image Intensification*, Technical Note, Soesterberg, Netherlands: Institute for Perception RVO-TNO, National Defense Research Council, 1973.

[50] Granger, E. M., and K. N. Cupery, "An Optical Merit Function (SQF) which Correlates with Subjective Image Judgements," *Photographic Science and Engineering*, Vol. 16, 1972, p. 221.

[51] Campbell, F. W., and J. G. Robson, "Application of Fourier Analysis to the Visibility of Gratings," *Journal of Physiology*, Vol. 197, 1983, pp. 551-566.

[52] Carlson, C., and R. Cohen, "A Simple Psychophysical Model for Predicting the Visibility of Displayed Information," *Proceedings of the Society for Information Display*, Vol. 21, 1980, pp. 229–245.

[53] Mitchel, R. H., *SAR Image Quality Analysis Model*, Ann Arbor, MI: Environmental Research Institute of Michigan, 1974.

[54] Rosell, F. A., and R. H. Willson, "Recent Psychophysical Experiments and the Display Signal-to-Noise Ratio Concept," in L. M. Biberman (ed.), *Perception of Displayed Information*, New York, NY: Plenum Press, 1973.

[55] Lubin, J., *A Methodology for Imaging System Design and Evaluation*, Princeton, NJ: David Sarnoff Research Center, 1995.

[56] Avadhanam, N., and R. Algazi, "Evaluation of a Human Vision System Based Image Fidelity Metric for Image Compression," *SPIE Conference on Applications of Digital Image Processing XXII*, Denver, Colorado, 1999, pp. 569–579.

[57] Nunes, P. R., A. Alcaim, and M. R. da Silva, "Compression of Satellite Images for Remote Sensing Applications," *Proceedings of the XVII Congress of the International Society for Photogrammetry and Remote Sensing*, Washington, DC, August 1992, pp. 479–483.

[58] Halford, C. E., et al., "Developing Operational Performance Metrics Using Image Comparison Metrics and the Concept of Degradation Space," *Optical Engineering*, Vol. 38, No. 5, pp. 836–844.

7

Target and Environmental Considerations

In this chapter, we begin an analysis of factors in the image chain (Figure 7.1) as they affect performance prediction and modeling. We begin with the target. Emphasis is on spatial properties, although some discussion of wavelength-related signatures is provided. A discussion of deception and denial effects is included, although little material is available on the topic. Finally, a discussion of atmospheric models and environmental considerations is included.

7.1 Target Effects

A discussion of target effects begins with the findings of Johnson [1], which are shown in Table 6.2. Johnson's work dealt with image intensifiers used on the ground. Johnson thus used side views of vehicles. Several other investigators showed similar findings [2–4] but disagreements also arose. Wagner and van Meeteren argued that results were vehicle dependent and found 2:1 differences as a function of vehicle type [3]. Bennet, et al., although showing agreement for tactical vehicles, indicated that as many as 2400 lines were required for industrial target recognition [4]. Another study showed a recognition requirement of 7–12 lines for one level of recognition (vehicle type) and 5–8 lines for another (tracked versus wheeled) [5].[1]

---

1. Several of the studies dealing with number of subtending lines also varied scale or visual display angle. Here, we ignore that variable and treat it later as a display effect.

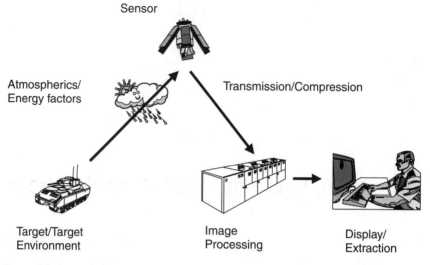

Sensor

Atmospherics/
Energy factors

Transmission/Compression

Target/Target
Environment

Image
Processing

Display/
Extraction

Figure 7.1 Image chain.

Rosell and Willson [6] performed a study using photographs (low oblique side views) of vehicles. They observed that each of the vehicles evaluated had an isolated characteristic feature and that "to recognize the object, the feature must be discerned." They also observed that the feature was about 1/8 the total minimum width² of the object. For identification, the needed features were closer to 1/13 the overall minimum width. Rosell and Willson summarized the Johnson criteria as requiring eight TV lines for recognition and 12.8 for identification. This suggests a rational basis for the Johnson criteria as well as an explanation of deviations from the criteria. For example, in a study using photos of vehicle models taken from above, as many as 18 scan lines across the width of a vehicle were required for tank identification [7]. Many of the early recognition studies appear to have ignored sampling effects in that a feature the size of a single line would be detectable only if in phase.

In a study using aircraft photos, the photos were captured on video and displayed with differing numbers of subtending lines [8]. Side, head-on, and oblique photos of six jet fighter aircrafts were used. The displayed aircraft subtended 6.0, 10.2, and 14.4 minutes of arc (minimum dimension), and the number of scan lines was 7.2, 10.1, or 14.4 across the minimum dimension. The observers' task was to identify the aircraft model.

---

2. In this case, minimum width was the dimension perpendicular to the line of sight.

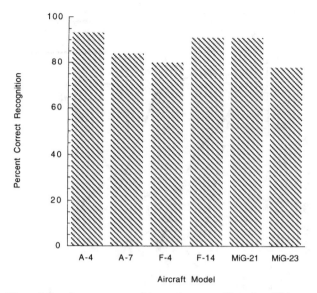

Figure 7.2 Effect of aircraft type on recognition performance. (Data from [8].)

Figure 7.2 shows results as a function of aircraft type for the largest number of subtending lines and visual angle, with results averaged across orientations. The effects of aircraft model are apparent. Figure 7.3 shows results as a function of orientation with results summed across visual angle and subtending line conditions. Performance on side views is better than on head-on and oblique, but there is a strong aircraft model interaction. Confusion matrices provided in the reference report as well as inspection of the aircraft photos suggest reasons for these results based on identifying features. For example, in a head-on view, both the A-4 and F-4 have a rounded engine intake on both sides of the fuselage. In 43% of the responses, the F-4 was misidentified as an A-4.

In another study of aircraft recognition, aircraft plan view silhouettes were viewed by subjects at varying distances to simulate differences in ground resolution [9]. Simulated ground resolution values ranged from 0.5 to 8 ft in $\log_2$ steps; a value of 12 ft was also used. Subjects reported their ability to correctly classify aircraft identification features such as wing shape, fuselage shape, etc. A total of 16 aircraft silhouettes were used and observers were asked to classify 13 different feature categories. Each aircraft was then defined in terms of the minimum number of features needed for identification.

The silhouettes were analyzed to define the identification features for each aircraft. The F-111, for example, could be distinguished from other aircraft by the shape of the leading edge of the wing. The F-100 required

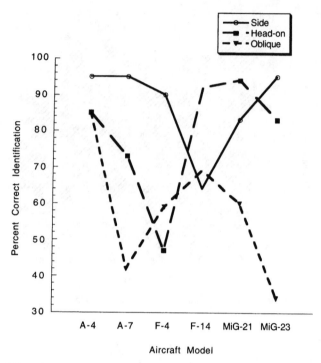

Figure 7.3 Effect of aircraft orientation on recognition performance. (Data from [8].)

identification of the shape of the forward portion of the fuselage plus the location of the engine intake ducts. Simulated ground resolutions required to identify the aircraft on the basis of these distinguishing cues ranged from 0.5 to 12 ft, a 24:1 range. Overall target size accounted for less than 50% of the variance in ground resolution required to distinguish identifying cues. A subsequent study was performed using aerial photos of aircraft (vertical photos) displayed at varying viewing distances to again simulate ground resolution differences [10]. Results were compared to those from the previous study. In 70% of the comparisons made, results (in terms of required ground resolution) were the same. In 28% of the cases, the photo required more resolution (generally one step or twice the resolution). No attempt was made to control or normalize contrast and SNR was not controlled.

In a study of the effects of ground resolution [11], analysts were asked to perform a series of detail analysis tasks on industrial and airfield scenes. Ground resolutions of the photos ranged from 3 to 30 inches for the industrial scene and 6 to 30 inches for the airfield scene. The cues required to perform the tasks were subsequently defined by experienced analysts and the relationship between cue size and performance defined as a function of

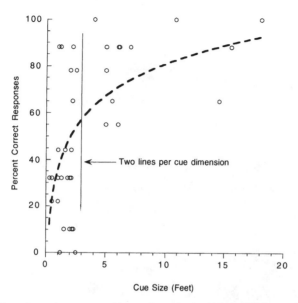

Figure 7.4 Effect of cue size on recognition performance for 30-inch GRD image. (Data from [9].)

ground resolution. Results are shown in Figure 7.4 for the airfield scene and the 30-inch ground resolution image. Analysis was performed to define the cue size producing the maximum value of chi square for each scene/GRD combination (12 and 30 inch only) and from this, the number of subtending resolution lines.[3] Results ranged from 1.2 to 2.4 lines. These results are consistent with those from Rosell and Willson [6] but extend the results to a different class of objects. Note that these findings are also consistent with Johnson's detection criterion in that two lines are required to detect the identifying feature or cue. We emphasize again, however, that contrast and signal-to-noise (S/N) must be sufficient. We conclude that resolution, defined in lines on the target, must be sufficient such that required recognition or identification features are subtended by at least two lines. Although the Johnson criteria are reasonable for low aspect angle side views of vehicles, they are not necessarily correct for other views and types of targets.

A series of photos and figures illustrate resolution issues. Figure 7.5 shows side views of military vehicles at four resolutions. The Johnson criterion for tank recognition is 6.3 line pairs, or 12.6 lines.[4] For the side view, this criterion seems reasonable, although in the example, identification

---

3. At 30-inch GRD, a resolution line was 15 inches.
4. Averaged across three models of tanks.

37 Lines

18.5 Lines

9.3 Lines

4.7 Lines

Figure 7.5 Side view of tanks at four resolutions.

appears possible at 9.3 lines. The recognition criterion is 6.9 lines; again reasonable in terms of the illustration.

Figure 7.6 illustrates the same tanks from the top. Here, 9.3 lines are sufficient for identification and 4.7 lines are barely sufficient for detection. Note that the background adds to the difficulty in detection. In Figure 7.7, a tank and a truck are shown at four resolutions. Even at the lowest resolution (4.7 lines), the difference between the two is recognizable.

Figure 7.8 shows four aircraft at three resolutions. At 18.5 lines across the minimum dimension (wingspan), the aircraft model cannot be identified. At 9.3 lines, the aircraft are barely detectable.

In the previous discussion, only the visible portion of the spectrum was considered. The same rules apply to IR, however, at the feature level, sometimes called the signature level. A signature is a set of unique recognition cues or features. Thermally active targets may exhibit unique signatures related to,

37 Lines

18.5 Lines

9.3 Lines

4.7 Lines

Figure 7.6 Plan view of tanks at four resolutions.

for example, the location of engine vents and exhausts. Figure 7.9 provides a hypothetical example. In the visible realm at full resolution of 37 lines/width [Figure 7.9 (left)], the two tanks are distinguished on the basis of turret shape differences, hatch differences, the presence of a machine gun cupola, and deck differences. When resolution is reduced to 9.3 lines per width [Figure 7.9 (middle)], the two vehicles can no longer be distinguished. In addition, compare this figure with Figure 7.6, where the contrast difference and the presence of shadows make a significant difference in the ability to perform

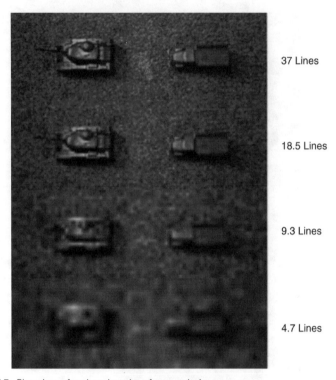

37 Lines

18.5 Lines

9.3 Lines

4.7 Lines

Figure 7.7 Plan view of tank and truck at four resolutions.

identification. In Figure 7.9 (right), engine hatches are shown with a simulated IR signature. The difference between the two signatures is distinctive, even though the remainder of the features cannot be distinguished. If we assume a width of 12 ft for the tanks, one line equals 1.3 ft. If the task was to identify moving tanks (engines on), a GSD of 1.3 ft might suffice. Identifying tanks in garrison would require an estimated 2.6-ft GSD (assuming inactive engines). A similar situation exists with aircraft. Engines and engine exhausts may be evident and serve as identifying cues. Identification can thus occur at coarser resolutions than would otherwise be required.

Given a recognition or identification problem, there are two approaches to defining required resolution (in terms of number of subtending lines). The first is to define the recognition features or cues and then to define their sizes. A minimum of two resolution lines should subtend the minimum size feature. The recognition features can be defined from recognition keys, by inspection, or graphically. For example, Figure 7.10 shows plan view silhouettes of two aircraft differing in length and wingspan by less than 2 ft. Obvious

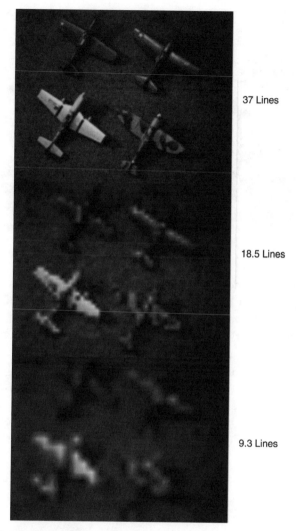

37 Lines

18.5 Lines

9.3 Lines

Figure 7.8 Plan view of aircraft at three resolutions.

differences include aft wing shape (a), fuselage shape (b), and horizontal stabilizer shape (c). The wing and stabilizer shape differences also result in a significant difference in the distance between the aft edge of the wing and the forward edge of the horizontal stabilizer. More subtle differences include wing tip shape (d), location of pitot tube (e), and horizontal stabilizer tip shape (f).

A crude scaling of the silhouettes indicates that the width of the cues ranges from less than 1 ft (pitot tube, e) to 4 ft (aft fuselage shape, b). The

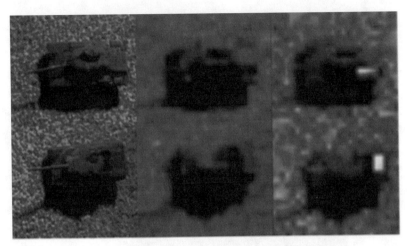

Figure 7.9 Effects of shadow and simulated IR signature on tank identification.

difference in the distance between the wing and stabilizer is about 3 ft. Assuming the target will be seen in plan view, a 2-ft GSD would suffice, although a 1.5-ft GSD would provide two cues. Note that this is a very simple example; actual imagery would show other potential cues. For example, the aircraft on the top has a high mounted wing and stabilizer, the one on the bottom has a low mounted wing and stabilizer. There is roughly a 4-ft difference relative to the bottom of the fuselage that would be evident on an oblique view; shadow differences might also be apparent.

A second method of defining requirements involves use of the visual system as a detector. The eye can resolve 10 line pairs per millimeter at a 10-

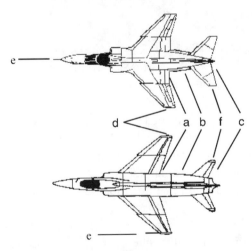

Figure 7.10 Cue differences for two aircraft silhouettes.

inch viewing distance [12]. Knowing the scale of a drawing or image, one can view the drawings or images at a distance such that differences between the two cannot be resolved. One moves closer until the differences can be resolved and computes the required GSD on the assumption that the resolution of the eye scales to GSD. This was shown to be approximately true in an earlier referenced study [9].

Equivalent ground resolution (in feet) is calculated as

$$EGR = \frac{Scale\ Factor(Viewing\ Distance/10)}{10(304.8)} \qquad (7.1)$$

where scale factor is the reciprocal of the image scale (image scale is the ratio of image to ground distance, for example, 1/1000), 10 in the denominator is the resolution of the eye in line pairs/mm at a 10-inch viewing distance (the 10 in the numerator), and 304.8 is the conversion of millimeters to feet. The conversion of equivalent ground resolution to equivalent GSD (EGSD) is approximated by

$$EGSD = \frac{EGR}{2.8} \qquad (7.2)$$

where EGR is equivalent ground resolution and the factor of 2.8 (rather than 2) accounts for sampling effects using the Kel factor of 0.7. For the example shown, the required EGSD was estimated at 2.8 ft by the author.

In the case of radar, man-made targets are defined by a series of point returns. These returns, and thus the appearance of a target, vary as a function of imaging geometry. Furthermore, because returns are specular, there is not a clear relationship between object size and intensity returns or signatures. Radar returns may come from surfaces that reflect single, double (dihedral), and triple (trihedral) bounces of radar energy. In some cases (e.g., cavity returns) many multiple bounces occur. Cavity returns come from open cylinders. Objects with many reflecting surfaces are more likely to provide unique signatures than those with few reflecting surfaces. Also, because radar provides its own source of energy and does not penetrate objects, radar shadows can be a significant signature component. Shadows, of course, vary with imaging geometry. The target signature thus can vary substantially as a function of orientation and look angle. For these reasons, there is not a clear relationship between SAR impulse response (resolution) and recognition or identification performance.

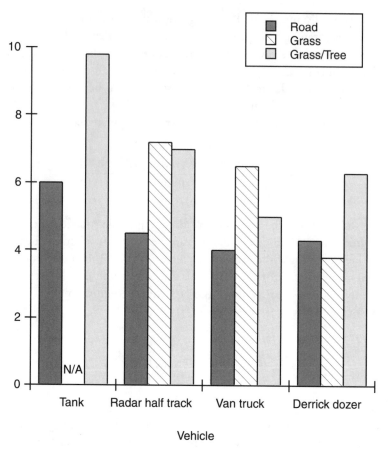

Figure 7.11 Effect of background on vehicle identification. (Data from [6].)

With a few exceptions [8,10] the data presented thus far dealt primarily with vehicles on uniform backgrounds and generally high contrast (relative to target returns) backgrounds. A few studies have investigated the effect of the background on performance. Rosell and Willson placed model photos on a scene photo to study the effect of background [6]. Figure 7.11 shows the results. Targets were of a uniform size (eight lines across the minimum dimension). The authors conclude that had size been varied, SNR_D would have been constant.[5] They define SNR_D thresholds for recognition as shown in Table 7.1.

In another study of vehicle identification as a function of background, models were placed on a terrain model, photographed, and displayed through

---

5. See (6.26).

Table 7.1
SNR_D Recognition Thresholds

Discrimination level	Background*	TV lines	Threshold SNR_D**
Detection	Uniform	1	2.8
Detection	Clutter	2	2.5
Recognition	Uniform	8	2.5
Recognition	Clutter	8	3.4
Identification	Uniform	13	3.0

*TV lines per minimum dimension.
**Single bar with spatial frequency of 500 lines per picture height.

a video system [2]. Results are summarized in Figure 7.12. Performance was lower on the sand background than the foliage background because of a contrast problem. The models were apparently rather specular and vehicles showed strong highlights. The highlights reduced the target contrast against the bright sand background. For the foliage background, there was about a

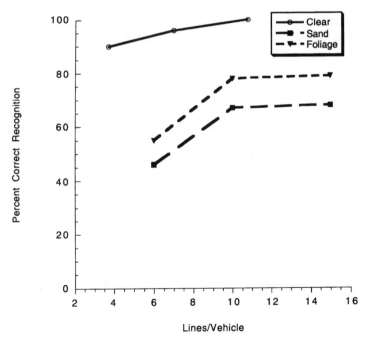

Figure 7.12 Effect of background on vehicle identification. (Data from [2].)

25% drop in performance relative to the clear background where performance appears to reach an asymptote at 10 lines per vehicle. Performance did not improve as the number of scan lines increased.

Based on the studies discussed in this section, we conclude that the effect of background in the visible spectrum can be quite significant, at least for situations where the recognition cues are on the periphery of the object and are not hidden. The effect of background also varies as a function of imaging geometry. If the geometry is such that the contrast of the distinguishing cues is affected by the background (as it was in [2]), the effect is likely to be greater than when the cues are defined by internal (to the target) contrast. Similarly, lighting effects can be significant in terms of the effect on target and feature contrast.

In the IR and radar realm, the effects may be even greater. For inactive (no internal source of radiant energy), there is theoretically a crossover point for some targets where they radiate the same level of energy as a background. Objects absorb energy from the sun during daylight and emit energy at night. Man-made objects in particular tend to radiate more energy than their background during the day and less at night. At some point, then, in both the evening and morning, a homogeneous object and its background show no contrast.

In the case of radar, the recognition task entails separation of point returns from the target background. Target recognition in high clutter backgrounds is thus more difficult than in low clutter backgrounds. The relationship between clutter levels and recognition performance has not yet been defined.

7.2   Deception and Denial

Deception and denial (D&D), also known as camouflage, concealment, and deception (CC&D), has been practiced at least since the American Revolutionary War. Rather than fighting in open formations and bright uniforms, some American soldiers fought in drab uniforms from the concealment of trees. Vegetation and camouflage continue to be used for deception and denial.

Deception is any act that deceives or fools an adversary. In the context of this book, deception includes the use of camouflage and false or dummy targets. Camouflage involves disguising the appearance of an object such that it cannot be distinguished from its (generally natural) surroundings or can not otherwise be recognized as a target. Dummy targets or decoys can assume many forms, from crude to highly realistic. Even natural objects or features can serve as unintended decoys. Tree shadows along a road can look like a column of vehicles, for example [13]. The success of a decoy is a function of the time and money available for construction and the quality of the resources expended to detect the presence of a decoy. An aircraft dummy can be constructed from wood and fabric or from metals and plastics. The greater the fidelity to a real aircraft, the greater the construction expense and the greater the resources needed for detection.

The availability of multispectral and multisensor systems has perhaps made the detection of dummy targets easier. To be effective against a multispectral system, the dummy must exhibit the same reflective and emissive properties as the real object in each spectral band. As the number of bands increases, so does the difficulty of constructing an accurate dummy. Similarly, the availability of a visible, IR, and SAR capability requires much more effort to defeat than for any single sensor. Temporal and aspect angle diversity also provide capabilities for detecting dummies and decoys. An IR dummy, for example, may have different emissivity and thermal mass than the real object and thus cools or warms up at a different rate. Aspect angle diversity is particularly useful in the SAR case, because a dummy would need to have the same signature at all aspect angles as the real object in order to be completely successful.

Camouflage includes the use of paint, camouflage nets, and natural cut vegetation. Paint (or fabric dyes in the case of uniforms) is used to make the target less distinct with respect to common backgrounds and to disrupt the outline of the target. NATO vehicles are painted with a three-color pattern as shown in Figure 7.13 [14]. The colors used are varied depending on

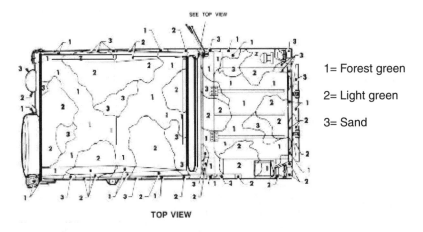

1= Forest green

2= Light green

3= Sand

TOP VIEW

Figure 7.13 Camouflage pattern for top of jeep (summer verdant).

climate and vegetation. From a sufficient distance, the target blends in with its background. Further, the disruptive paint pattern makes the characteristic linear features of man-made objects more difficult to separate from the background. Although there has been considerable research in developing metrics for target distinctiveness, it does not yet appear to be of value in designing or operating an S&R system. The difficulty arises in the large variability in target distinctiveness that can result from small changes in camouflage procedures and target location.

A review of a study on distinctiveness metrics is illustrative [15]. In the study, a series of metrics were applied to measure target contrast, target texture difference, and boundary strength. Contrast metrics included the area weighted contrast (ΔT), the standard deviation of the target and background fields (Doyle metric), and the number of pixels in the target differing from the background by more than two standard deviations of the background distribution (Eff_POT). Texture distinctiveness was measured by the average co-occurrence error (ACE). Boundary strength was measured using the average contrast along the target boundary (average boundary strength, ABS) and the ratio of the average contrast along the target-background boundary to the average contrast between adjacent pixels in the vicinity (ratio average boundary strength [RABS]).

The targets were natural patterns embedded in other natural patterns. Observers judged target distinctiveness in a paired comparison study. Using a comparative scaling technique, the data was assigned a scale ranking from 0 to 4. Correlations with the distinctiveness metrics ranged from 0.136 (ΔT) to 0.825 (ACE). Using the data published in [15], we performed a stepwise multiple regression on the data using one of each type of metric. The resulting regression equation predicting perceived target distinctiveness was

$$D = -1.492 + 1.381 * ACE + 0.361 * RABS \tag{7.3}$$

ACE is defined as

$$ACE = \frac{1}{\mathcal{F}_{NGLC}} \sum_{\Delta \in \mathcal{D}} \sum_{i=0}^{G-1} \sum_{j=0}^{G-1} |P_t(i, j | \Delta) - P_b(i, j | \Delta)| \tag{7.4}$$

and $\mathcal{F}_{NGLC}$ is the number of vectors in set $\mathcal{D}$, G is the number of gray levels, $|P_t(i, j | \Delta)$ is the joint probability a pixel of gray level I and a pixel of gray level j given $\Delta = [\Delta_x \Delta_y]$ for the target pattern, and $P_b(i, j | \Delta)|$ is the joint probability for the background.

RABS is defined as

$$\text{RABS} = \frac{\text{ABS}}{(1/n_{region})\Sigma_{i=1}^{n_{region}} c(i)} \tag{7.5}$$

where $c(i)$ is contrast and n_{region} is the number of pixel pairs in the target field or in the background near the target. ABS is simply the average boundary strength for the target.

Although the ACE metric accounted for most of the variance, the RABS metric contributed significantly to the prediction equation. The value of R^2 for (7.3) was 0.72 and the standard error of prediction was 0.52 (on a scale with a range 0–4). It appears that this metric could be used to assess, on a relative basis, the effects of a particular camouflage situation or procedure. The difficulty with such an approach, however, is relating the metric to actual performance.

Camouflage can also consist of camouflage netting designed to cover an object and make it blend into the background. Ideally, the camouflage netting should not be detectable, thus hiding the presence of an adversary. White camouflage would be used in a winter environment, and camouflage resembling vegetation would be used in a temperate environment. Generally, the camouflage netting is opaque if properly applied. In other words, even if the netting is detected, what is underneath the netting is not identifiable. In that sense, then, the netting may be considered a form of denial. In a similar sense, hiding objects along tree lines and under tree cover is a form of both deception and denial (or concealment). Finally, hiding activity by smoke or darkness constitutes denial.

Again, the best solution to camouflage (as well for concealment and denial) may be the use of multisensor and, perhaps to a lesser degree, multispectral approaches. SAR and IR (sometimes) can defeat denial by smoke or darkness. For example, a hot emitter may be detectable in tree cover even when the overall object is not detectable. In order to be effective against detection, camouflage must show the same spectral response as natural objects. Color IR film was developed to detect the difference between natural vegetation and dead vegetation or camouflage netting. Multispectral systems can detect camouflage (paint, netting, cut vegetation) that does not show the same spectral properties as natural vegetation. For a single sensor, however, there is no good rule of thumb defining design approaches to defeating deception and denial. At its best, camouflage can defeat even close observation. Consequently, there does not appear at present to be a resolution or sensitivity trade space for broadband systems. Increased resolution and sensitivity may help, but it also can be defeated.

7.3 Atmospheric Effects

The atmosphere affects the transfer of energy from the source to the target and from the target to the sensor. In this discussion, we also include the effects of the angular relationship between source and target.

For many applications, knowledge of energy transfer is required for sensor calibration. The goal is to define the energy emitted or reflected by objects in the scene in absolute terms (e.g., temperature and emissivity for an IR system). In the context of S&R modeling, however, the goal is to predict the effect of the atmosphere on target-to-background contrast (or signal-to-noise ratio), and accuracy becomes somewhat less important.

In the 1970s, the Air Force began to develop atmospheric radiation transfer algorithms for use in computer modeling [16]. The first result of this effort was released in 1972 as LOWTRAN, the Low Resolution Transfer Code [17]. Subsequent work has resulted in MODTRAN (Moderate Resolution Transfer Code), and FASCODE (Fast Atmospheric Signature Code) [18,19]. Development is continuous. MODTRAN code can be found at http://imkpcdemo.fzk.de/isys-public/. It is also available with a PC interface from http://www.ontar.com.

MODTRAN is used to calculate the transmission and radiance that would be observed by a sensor in the wavelengths between the UV and microwave looking through the atmosphere from a given geometry. The parameters include:

- Ground emissivity;
- Ground height;
- Ground brightness temperature;
- Humidity profile;
- Temperature profile;
- Scan angle.

For a given wavelength window, radiance is computed for each of n bands. These values must then be integrated over the response window and sensitivity of the sensor to compute the radiance at the sensor. An example is provided in Section 10.1.1. MODTRAN is used in computing a signal-to-noise ratio.

MODTRAN assumes the atmosphere consists of multiple layers, each layer having a temperature and a distribution of gases, water vapor, and aerosols. Distribution assumptions are based on user-supplied data concerning the air mass type (e.g., maritime, continental), season (e.g., mid-latitude summer, winter), and meteorological visibility (e.g., 20 km, rural haze).

MODTRAN computes radiance through the atmosphere, but does not handle other effects related to the relative position of the energy source and sensor. For sensors where the sun is the primary energy source, the sun angle is the angle between the ground plane and the sun (Figure 7.14). Sun angle controls the length of shadows and also affects the length of the atmospheric path. Sun azimuth is the horizontal angle to the sun. The combination of sun azimuth and target orientation affects the usefulness of a shadow in aiding detection or recognition. Figure 7.15 shows an example. The shadow in Figure 7.15(b) enhances both contrast and recognition. As shadow angles change, the usefulness of the shadow varies. Very long shadows can interfere with other objects in the scene and may be difficult to interpret; non-orthogonal shadows can also be more difficult to interpret. This effect is generally not captured in performance models as a separate term because of the difficulty of a priori prediction.

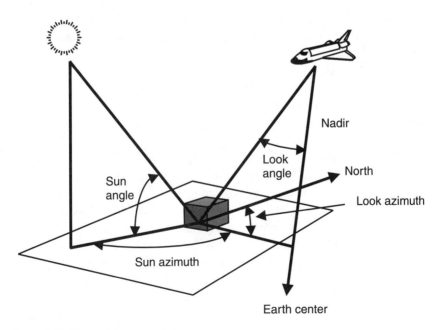

Figure 7.14 Sun angle measurements.

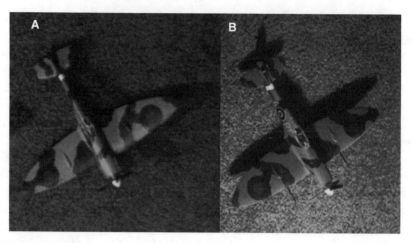

Figure 7.15   Effects of shadows on recognition.

7.4 Summary

This chapter discussed the effects of target and backgrounds and the MODTRAN atmospheric model. Target recognition and identification are achieved by detecting differences between like objects. These differences are called recognition cues. If the cues cannot be detected or resolved, the recognition/identification task cannot be successfully performed. The Johnson criteria defined resolution requirements for specific viewing conditions and recognition levels for selected targets. These requirements have been validated for the same viewing conditions in other studies, but generally not for other targets and viewing conditions. It was demonstrated that if the required recognition cue is subtended by two resolution lines (GSD is half the size of the cue), recognition can occur. Techniques for defining cues and required GSD were provided.

Deception and denial techniques increase the difficulty of target detection and identification. Although models exist to classify relative target difficulty related to camouflage, no single sensor-specific performance prediction models are available. The success of deception and denial techniques is only resource-limited. Each improvement in the S&R system can be countered with improved D&D techniques, at least in a static situation.

The atmosphere intervenes between the target and the sensor and has the effect of reducing target contrast. MODTRAN is a model used to calculate the effects of the atmosphere on radiance reaching the sensor. The relative position of the energy source, as well as the target orientation relative to the source, can also affect recognition performance.

References

[1] Johnson, J., "Analysis of Image Forming Systems," *Proceedings of Image Intensifier Symposium,* Ft. Belvoir, Virginia, 1959, pp. 269–273.

[2] Erickson, R. A., and J. C. Hemingway, *Image Identification on Television,* NWC TP 5025, China Lake, CA: Naval Weapons Center, 1970.

[3] Wagenaar, W. A., and A. van Meeteren, *Discrimination and Identification in Noisy Line- Scan Pictures,* RN-12F-1967-16, Soesterberg, Netherlands: Institute for Perception RVO-TNO, National Defense Research Council, 1967.

[4] Bennett, C. A., et al., *A Study of Image Qualities and Speeded Intrinsic Target Recognition,* Owego, NY: IBM Federal Systems Division, 1963.

[5] Scott, F., P. A. Hollanda, and A. Harabedian, "The Informative Value of Sampled Images as a Function of the Number of Scans per Scene Object," *Photographic Science and Engineering,* Vol. 14, No. 1, 1970, pp. 21–27.

[6] Rosell, F. A., and R. H. Willson, "Recent Psychophysical Experiments and the Display Signal-to-Noise Ratio Concept," in L. M. Biberman (ed.), *Perception of Displayed Information,* New York, NY: Plenum Press, 1973.

[7] Hollanda, P. A., and A. Scott, "The Informative Value of Sampled Images as a Function of the Number of Scans per Scene Object and the Signal-to-Noise Ratio," *Photographic Science and Engineering,* Vol. 14, No. 6, 1970, pp. 407–413.

[8] Lacey, L. A., *Effect of Raster Scan Lines, Image Subtense, and Target Orientation on Aircraft Identification on Television,* China Lake, CA: Naval Weapons Center, 1975.

[9] Jones, R., J. C. Leachtenauer, and C. Pyle, *Critical Cue Requirements Studies: Aircraft Feature Recognition,* D180-10557-1, Kent, WA: Boeing Co., 1970.

[10] Leachtenauer, J., and R. Jones, *Critical Cue Requirements Studies: Photo Validation of Aircraft Feature Recognition,* D180-10557-2, Kent, WA: Boeing Co., 1971.

[11] Leachtenauer, J. C., and G. P. Boucek, *Ground Resolution Study,* D180-10557, Kent, WA: Boeing Co., 1970.

[12] Farrell, R. J., and J. M. Booth, *Design Handbook for Imagery Interpretation Equipment,* Seattle, WA: Boeing Aerospace Co., 1984.

[13] Martinek, H., and R. Sadacca, *Error Keys as Reference Aids in Image Interpretation,* TRN-153, Washington, DC: U.S. Army Personnel Research Office, 1965.

[14] U.S. Army, Color, Marking and Camouflage Painting of Military Vehicles. TB 43-0209.

[15] Copeland, A. C., and M. M. Trivedi, "Signature Strength Metrics for Camouflaged Targets Corresponding to Human Perceptual Cues," *Opt. Eng.,* Vol. 37, No. 2, pp. 582–591.

[16] Anderson, G. P., et al., "FASCODE/MODTRAN/LOWTRAN: Past/Present/ Future," http://imkpcdemo.fzk.de/isys-public/Software-tools/Modtran/science/fa-mo-lo.htm, 1996.

[17] Kneizys, F. X., *Users Guide to LOWTRAN 7*, AGGL-TR-88-0177, Hanscom Air Force Base, MA: Air Force Geophysics Laboratory, 1988.

[18] Berk, A., L. S. Bernstein, and D. C. Robertson, *MODTRAN, A Moderate Resolution Model for LOWTRAN 7*, GL-TR-89-0122, Hanscom Air Force Base, MA: Air Force Geophysics Laboratory, 1989.

[19] Smith, H. J. P., *FASCODE-Fast Atmospheric Signature Code (Spectral Transmittance and Radiance)*, AFGL-TR-78-0081, Hanscom Air Force Base, MA: Air Force Geophysics Laboratory, 1978.

[20] Maver, L., et al., "Aerial Imaging Systems," in J. V. Sturge, V. Walworth, and A. Shepp (eds.), *Imaging Processes and Materials*, New York, NY: Van Nostrand Reinhold, 1989.

8

Image Processing Considerations

In this chapter, we examine the effects of image processing on S&R system modeling and performance prediction. We begin with the effects of bandwidth compression (BWC). The goal is to demonstrate the effect of BWC as a component in the image chain. We follow this with discussions of the effects of image enhancement processing. At this stage of the image chain, an image has been collected and processed for viewing, but has not yet been displayed. In the following chapter, we examine the contribution of the display and observer.

8.1 Bandwidth Compression

Digital imagery requires extensive communication bandwidth to pass data from sensors to users and from one user to another. A single frame of Corona imagery, when digitized at full resolution, is 2.6 gigabytes (GB) of data [1]. This data fills four 650-megabyte (MB) compact discs (CDs). Digitizing all of the domestic Corona KH-4A coverage (10 million nmi^2 and 6700 frames) would fill close to 27,000 CDs.

A single second of digitized Predator video provides ~26 MB of data [2]. A 1-hour mission fills roughly 144 CDs. A 20-GB hard drive could hold less than 2 minutes of video data. Full digital formats provide up to 116 MBps of data (~18 CDs per second).

A T-1 link carries 1.54 megabits per second (Mbps) or 0.187 MBps. This equates to four frames (of 26-MB video) per second. It requires 139 hours to transmit 1 hour's worth of Predator video. A T-3 link carries 45

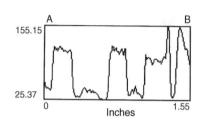

Figure 8.1 Pixel intensity trace (A to B) showing intensity correlation.

Mbps and requires ~4 hours to transmit 1 hour of Predator data. A phone line is theoretically capable of 56 kilobits per second (0.007 MBps). At this rate, slightly over 1 hour is needed to transmit a 1-second video clip. The U.S. Department of Defense (DoD) mandated common data link (CDL) carries a maximum of 274 Mbps and can thus handle Predator data in real time [3]. It cannot handle uncompressed full digital video formats, a little over 3 hours is needed to handle 1 hour of data.

The need for BWC is apparent. A variety of image bandwidth compression techniques exist. The goal in BWC is to maximize file compression and minimize information loss. Broadly speaking, compression uses redundancy to reduce file size. In a visible image, the same or similar gray levels occur in strings or groups (Figure 8.1). By recording only changes (and the magnitude of the changes), file size can be reduced. A technique called differential pulse code modulation (DPCM) uses this technique. Moving Pictures Experts Group (MPEG)[1] compression is used for video and capitalizes on frame-to-frame redundancy. It is not necessary to fully encode every frame; only the changes are important. The eye is less sensitive to color than luminance; file sizes can be reduced by reducing the color information pro-

---

1. The Motion Pictures Experts Group (similar to the Joint Photographic Experts Group, JPEG) began in 1988 and has developed syntaxes to standardize the coding of various video and audio input formats. A group called the Video Quality Experts Group (VQEG) is currently engaged in the development of video compression quality tools.

vided. Commercial video in the United States (National Television Systems Committee, NTSC) encodes the color data at half the resolution of the luminance data. Multispectral and hyperspectral imagery shows a good deal of correlation across or among bands; this correlation can be used to reduce file sizes. Pixel averaging or deletion is still another BWC technique. Finally, transforms into frequency space and compression in this space is another common technique.

There are as yet no models that accurately describe the effects of BWC on image interpretability or even subjective quality. It is common to express the effects of BWC in terms of statistical image differences such as PSNR and RMSE. PSNR is the peak signal-to-noise ratio as defined in (6.37); RMSE is the root-mean-square error as defined in (6.36). Unfortunately, these metrics do not always predict subjective image quality ratings or interpretability losses. The effects of bandwidth compression on subjective quality and interpretability vary as a function of compression type and rate, and image content. Compression can produce a variety of artifacts such as blockiness, noise, ringing, ghosting, jaggedness, and blurring. These artifacts do not all react in the same manner to statistical difference measures. Figure 8.2 provides an example. Three BWC schemes (one at two rates) were used to encode a picture [4]. The subjective quality of the compressed images was

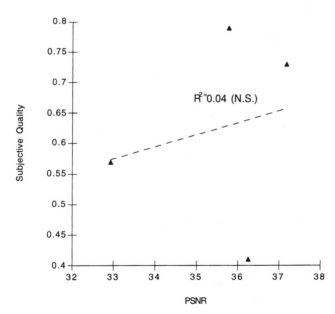

Figure 8.2 PSNR versus subjective quality ratings. (Data from [4].)

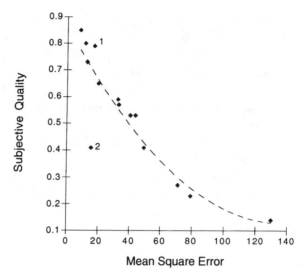

Figure 8.3 Mean square error versus subjective quality ratings. (Data from [5].)

rated by 39 observers relative to the original. A 10-point scale was used with subsequent normalization and transform to a 0 to 1 scale. The correlation between subjective quality and PSNR is low and not statistically significant. The difficulty is that PSNR does not respond strongly to low-contrast blockiness, which occurred with two of the four algorithms.

Somewhat better results were shown in a second study [5]; results are shown in Figure 8.3. Although the correlation between MSE and subjective quality is significant ($R = 0.9$), there are some discrepancies. Points 1 and 2 in Figure 8.3 have about the same MSE, but differ substantially in subjective quality scores. When only the four JPEG[2] data points are analyzed, the correlation between subjective quality and MSE increases to 0.99. Statistical measures may thus be useful for assessing a single compression type, but they are less useful when multiple compressions are evaluated in the same study.

8.1.1 Single Image Monochrome Compression

For still differential (nonvideo) monochrome imagery, three compression schemes are most common: pulse code modulation (DPCM), the discrete cosine transform (DCT), and wavelet compression (also known as the discrete

---

2. Joint Photographic Experts Group.

wavelet transform, DWT). These compressions may also be combined with other techniques such as resolution degradation.

DPCM is typically used for visually lossless compression. To produce visually lossless compression, compression rates of less than 2.5:1 are needed. DPCM encodes an image in blocks and takes advantage of tonal redundancy. Tonal (bit level) differences are quantized, and a table of those differences and the direction of the differences is used to encode the image and decode the compressed image. The differences are normally smaller than the actual values and, thus, space is saved. In an 8-bit image, 8 bits would be needed to encode an area of full white in the image. In a string of full white pixels without encoding, each pixel would require 8 bits to encode. If only differences are encoded, far fewer bits are needed, because the magnitude of the change is typically small until an edge is reached. Further reduction can be achieved by quantizing the difference table—not all possible differences are precisely encoded. Referring to Figure 8.1, the trace across the image is 204 pixels long and has an average count of 135. The largest count difference, however, is on the order of 30. Figure 8.4 shows the same image as Figure 8.1, but the image has been reduced to a 5-bit image and then expanded to 8 bits. Only 64 unique values exist in the image. The image histogram is also shown in the figure. The differences in the images are minor, suggesting that the accuracy of encoding differences need not be all that great.

DCT methods employ a discrete cosine transform, Huffman encoding, and quantization. With the transform to frequency space, higher frequency

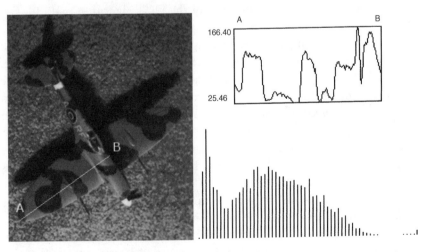

Figure 8.4 Five-bit image expanded to 8 bits. An intensity trace and histogram are also shown. Compare to Figure 8.1.

information can be discarded, if desired, thus saving space. The DCT transform is defined as

$$B(k_1 k_2) = \sum_{i=0}^{N_1-1} \sum_{j=0}^{N_2-1} 4A(ij)\cos\left[\frac{\pi k_1}{2N_1}(2i+1)\right]$$

$$\times \cos\left[\frac{\pi k_2}{2N_2}(2j+1)\right] \qquad (8.1)$$

where N_1 and N_2 are the height and width of the input image in pixels, $B(k_1 k_2)$ is the DCT coefficient in row k_1 and column k_2 of the DCT matrix. The DCT operates on 8×8 pixel blocks, and the input is a matrix of 8-bit values. The lower frequencies are in the upper left of the matrix and the higher frequencies in the lower right. The higher frequencies can often be discarded with no apparent loss in quality.

Huffman encoding is a method of assigning digital values based on frequency of occurrence (of another value or event). Shorter bit patterns are used for more common values and longer patterns for less common values. Quantizing again involves grouping small differences together. Not only does quantizing reduce file size, but it also impacts the apparent quality loss. Quantizing tables can be matched to the characteristics of the imagery to achieve a higher quality result at the same bit level.

Perhaps the most common form of the DCT algorithm is JPEG. Defined by the Joint Photographic Experts Group [6], JPEG is a standard method of compressing imagery. The degree of compression achievable with JPEG is in part dependent on the frequency content of the original image. If there is little high-frequency content, substantial compression can be achieved. Figure 8.5 shows a portion of a digitized image in its original form and in a compressed version, where a 10× reduction in file size has been achieved. Figure 8.6 shows portions of the same images illustrating various types of errors resulting from JPEG compression. Figure 8.7 shows a Briggs target [7] compressed to 5.4× and 8.7×. Even at the 5.4× reduction, artifacts are beginning to show.

With original aerial imagery, compressions on the order of 2:1 can be achieved without any noticeable loss; compressions of 4:1 show a detectable loss in information [8]. In a study of the effects of compression on softcopy photogrammetry, it was concluded that compression ratios greater than 5:1 may distort the image and were thus not recommended for photogrammetry [9].

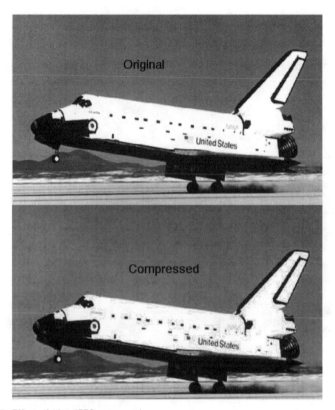

Figure 8.5 Effect of 10:1 JPEG compression.

Figure 8.8 shows data on the NIIRS loss expected for various DCT and DPCM compressions. Figure 8.9 shows the loss for NIMA Method 4 at much higher compression rates. NIMA Method 4 uses a combination of downsampling and JPEG compression to achieve low bit rates.

The data shown in the previous examples represents single compressions. It is not uncommon to compress and uncompress data several times in an image chain. Multiple compressions produce additional degradation. If a single JPEG compression at some defined rate produces a loss of NIIRS = X, multiple compressions will produce additional loss, even at the same rate. The problem is exacerbated by using different compression algorithms.

JPEG will be replaced by JPEG 2000. Again developed by a consortium of experts, JPEG 2000 employs wavelet technology. Wavelet compression operates in the frequency domain and evolved from short time Fourier transform analysis. Wavelet compression does not operate on small blocks, but rather treats the whole image and can thus take advantage of whole image correlation rather then just over small blocks. In addition to using DWT,

Original   Compressed

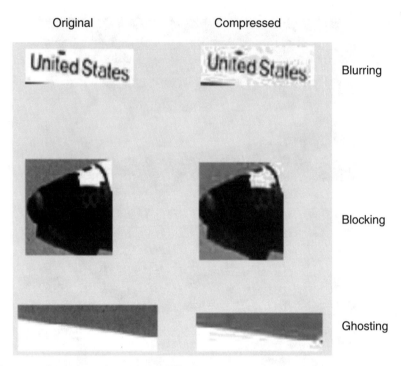

Blurring

Blocking

Ghosting

Figure 8.6 Enlarged portions of image in Figure 8.5 showing compression artifacts.

JPEG 2000 also uses embedded encoding. All versions of the compressed image are embedded in the bit stream such that any version up to the original encoded version can be transmitted or opened. If an image is compressed to, say, 4 bits/pixel, the file can be truncated to transmit only 1 bit/pixel. Alternatively, the whole file can be transmitted, but the receiver of the file may choose to open only a low-resolution version of the file. This capability has a further advantage in that it will not be necessary to combine or concatenate multiple different compressions. Currently, a compression occurs at a fixed rate. Files may be compressed and expanded multiple times. These multiple iterations produce quality degradation.

In a comparison of JPEG 2000 with current algorithms [8], significant improvements were seen in terms of NIIRS loss. Figure 8.10 shows the comparison. Lossless compression shows a 65% improvement with JPEG 2000; lossy compressions show a 20–60% improvement in throughput for the same bandwidth.

A second study compared three versions of a JPEG algorithm with JPEG 2000 [3]. JPEG algorithms included the baseline algorithm, an optimized JPEG, and NIMA Method 4. Results are summarized in Figure 8.11

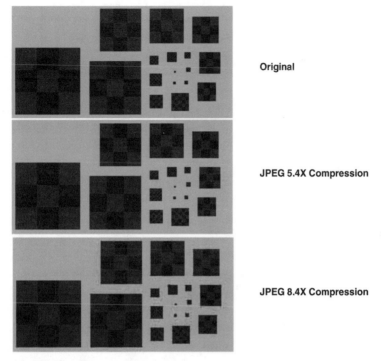

Original

JPEG 5.4X Compression

JPEG 8.4X Compression

Figure 8.7 Effect of JPEG compression on Briggs targets.

in terms of peak signal-to-noise ratio, mean square error, and mean absolute error. The study was performed at the two compression rates required to transmit the imagery within the capabilities of the CDL. The CDL has a maximum bandwidth of 274 Mbps with various combinations of 42.8- and 10.7-Mbps channels. There is no absolute correlation between NIIRS loss and any of the metrics used in the study. In a relative sense, however, it is apparent that JPEG 2000 shows better performance than the other algorithms. The effects of JPEG optimization can also be seen.

8.1.2 Multichannel Image Compression

For a single monochrome image, redundancy is a function of scene content and the nature of the imaging system (the degree of correlation in the spatial or frequency domain). With multichannel imagery (multispectral and hyperspectral), redundancy occurs in the spectral domain as well. A four-band, 8-bit per band multispectral image contains 32 bits per pixel. The IKONOS system collects four bands of multispectral data at 4-m GSD plus

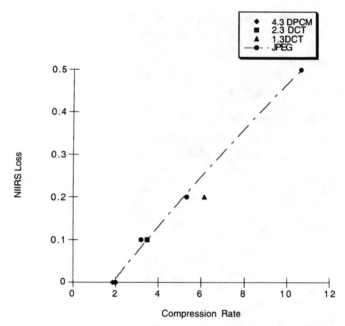

Figure 8.8   Effect of compression on NIIRS loss. (Data from [10].)

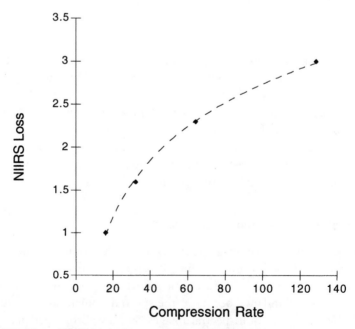

Figure 8.9   Effect of NIMA Method 4 compression on NIIRS loss. (Data from [8].)

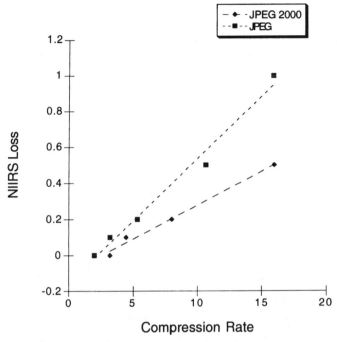

Figure 8.10 Comparison of wavelet (JPEG 2000) and DCT (JPEG) compression effects. (Data from [8].)

a 1-m GSD pan image, all at 11 bits per band per pixel. The cross-track swath width is 11 km; an 11-km square area contains 207 MB of data. Although the data is compressed to 2.6 bits per band per pixel for transmission to the ground, they are expanded for analysis and subsequent transmission.

For multispectral imagery (MSI) and hyperspectral imagery (HSI), the Karhunen-Loeve transform is applied [11]. This is based on a form of principal components analysis. It is a linear transform of a sample of points in *N*-dimensional space that defines a set of axes that best exhibit the properties of the sample. The principal axes include those where the variance of the sample is minimized. Given a sample of points (e.g., pixel intensity values in each of several bands), the center of gravity is defined and a dispersion matrix formed. The principal axes and the variance along those axes are defined in terms of eigenvectors and associated eigenvalues. Eigenvalues define an eigen image that can then be compressed using a transform in frequency space (e.g., JPEG). The effect is a reduction in the number of dimensions and a reduction in the intercorrelation among axes. A variation on this technique is used to perform spectral screening prior to the spectral decorrelation

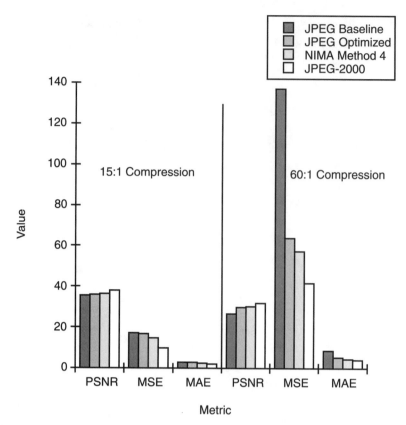

Figure 8.11 Comparison of JPEG algorithms and JPEG 2000. (Data from [3].)

process. The spectral screening identifies unique spectral signatures in a data block and preserves those signatures in the decorrelation process.

In a study of three-band composite MSI images, analysts were asked to rate the quality of compressed images relative to the uncompressed original. These ratings were then scaled to MS IIRS difference ratings [12]. At 1.2 bits per band per pixel (3.6 bits total), using the correlation among the bands reduced the impact of compression by ~0.2 MS IIRS units relative to a standard JPEG compression. This is a significant gain. As the number of bands increased, one would expect even larger savings.

8.1.3 Motion Imagery Compression

Video imagery offers both spectral and temporal redundancy. The substantial data rates of video (25–60 frames/second) require substantial compression, particularly as one moves to the tactical realm of limited bandwidth. A variety of compression techniques are commonly used with video, but only two use temporal redundancy.

The most common forms of commercial video are currently analog with NTSC (the U.S. standard) and PAL and SECAM (the European standards). The formats are interlaced; two fields make up a frame. Vertical resolution is defined by the number of active lines and ranges from 480 (NTSC) to 580 (PAL and SECAM). Horizontal resolution is a function of bandwidth, which ranges from 4.2 to 6.0 MHz. Common recording formats (VHS, 8 mm, Hi-8, S-VHS) apply spatial decimation to reduce required bandwidth. VHS, for example, reduces the number of lines to 240; Hi-8 uses 400 lines. The approach to spectral redundancy has been to define color as two luminance difference values (red and blue relative to total luminance with the green component being the unencoded remainder). Because the human visual system is more sensitive to luminance than color, the color difference signals are carried at half the bandwidth of the luminance signal. The work on video compression has occurred almost solely in the commercial entertainment industry. Preserving motion is considered very important. "Jerkiness" is considered very objectionable. The same criteria may not apply to S&R systems. Analog video is being replaced with digital video, which has higher quality and thus carries more data.

So-called motion JPEG is simply JPEG compression applied to video. There is no standard, and because temporal redundancy is not used, quality suffers relative to other forms. Motion JPEG has largely been replaced by some form of MPEG compression.

MPEG-1 was developed to compress full motion NTSC video to a fixed storage capacity, that of a CD-ROM. The required rate was 1.5 Mbps. MPEG-2 was developed to compress video down to a range of 3–10 Mbps. MPEG-2 exists in a variety of levels and profiles, as shown in Table 8.1 [13]. Levels indicate the file size; profiles denote the type of compression applied. MPEG-4 is being designed for higher compression rates.

In order to be transmitted and stored in a digital environment, analog video must be digitized. The conversion of analog to a meaningful digital file size is not straightforward. The resolution of NTSC video (the current U.S. standard for commercial TV) is defined as 483–490 lines of analog data, sometimes rounded down to 480 lines. The horizontal dimension is 1.33 times the vertical. The format could thus be envisioned as 483 by 642 (1.33 × 483). However, the horizontal dimension is analog and thus the true resolution depends on how it is sampled. The horizontal resolution is more accurately defined as

Table 8.1
MPEG-2 Levels and Profiles

	Profiles			
Levels	**Simple**	**Main**	**SNR**	**Spatial High**
High	1920 × 1152	1920 × 1152		
	60	60		
High-1440		1440 × 1152	1440 × 1152	1440 × 1152
		60	60	60
Main	720 × 576	720 × 576	720 × 576	720 × 576
	30	30	30	30
Low		352 × 288	352 × 288	
		30	30	

Note: Enhancement layer only.
Note: 1920 × 1152 indicates samples/line and lines/frame; 60 indicates frames/second.

$$R_H = \frac{[\text{Time (}\mu\text{sec) per active line} \times 2 \times \text{bandwidth (MHz)}}{\text{aspect ratio}} \qquad (8.2)$$

For NTSC video, the equation yields a value of 341 lines.

When analog data is digitized (D-1 format),[3] however, the signal is theoretically at 270 Mbps assuming 29.95 frames per second (the NTSC rate). If blanking information is eliminated and only active lines counted, the resolution is 704–720 pixels per line for 480–496 lines, or a rate on the order of 207 Mbps at 10 bits per pixel and 166 Mbps at 8 bits per pixel.

Analog video is also captured using VHS and 8-mm formats. These formats reduce the effective resolution to 240 lines by 352 pixels by sampling every other line and pixel (a 4:1 reduction). Super VHS and Hi-8 formats increase the number of lines to 400 or more and the number of pixels to 500 or more.

Digitizing analog video results in oversampling. The actual content of digitized NTSC video (4:2:0) is on the order of 55 Mbps (as opposed to 166–270). Digitized Hi-8 data is on the order of 30 Mbps.

In assessing the effects of compression, one must be careful in defining compression rates. For example, a compression of 8-mm video to a rate

---

3. The D-1 format is studio quality with CCIR 601 resolution of 720 × 480 × 29.95 fps.

of 6 Mbps can variously be considered a compression of 45:1, 28:1, or 5:1. The 5:1 compression rate is probably the most correct estimate, but the quality of the input data must also be considered. In the case of the 8-mm format, much of the high-frequency content has been lost in the acquisition process, making it easier to compress the remaining data.

Significant advances in video quality have been made with digital formats. Many of these formats use progressive scanning as opposed to interlaced scanning. In interlaced scanning, half the vertical lines are sent in one field and half in a second. If the rate is sufficient, no flicker is seen. With progressive scanning, the lines are sent a frame at a time. Table 8.2 lists digital video formats as defined by the DoD Video Working Group [14]. The formats are defined in terms of a video systems matrix (VSM).

To date, compression studies relating to S&R applications have been performed only with analog video as original input. In the first such study, data was captured by an 8-mm camera flown in a UAV and digitized [2]. The data (total of 30 clips at 22 seconds each) was digitized using a digital Betacam recorder to achieve rates of 6 Mbps, 1.54 Mbps, and 256 Kbps. Compression rates were defined in terms of a 166-Mbps input (D-1). A software version of MPEG-2 was used, and the clips at the two lowest rates were

Table 8.2
Digital Image Formats

Category	VSM level	Lines/samples	Frame rate (FPS)	Data rate
Ultra high	12–14	≥1920 × 1080P	50–120	TBD-3 Gbps
High	9–11	1280–1920 × 720–1080P	24–60	44.7 Mbps–4.5 Gbps
Enhanced	6–8	720–960 × 480–576P	50–60	5–270 Mbps
Standard	3–5	720 × 480–576P	24–30	3.8–270 Mbps
Low	2	352 × 480–576I or 240–288P	24–30	1.0–1.5 Mbps
	1	176 × 120–144P	10–15	256–768 Kbps
Very low	0	720–1920 × 480–1080	Still-2	56–512 Kbps

Note: Within categories of standard and above, levels differ in amount of bandwidth compression.
Note: VSM level 0 applies to products having spatial resolution of VSM 5/8/11 but very limited temporal resolution.
Note: P = progressive, I = interlaced.

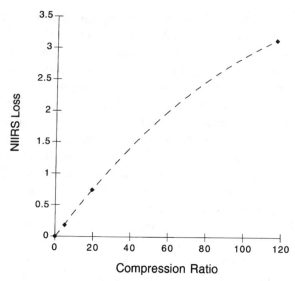

Figure 8.12   Effect of MPEG-2 compression on NIIRS loss. (Data from [2].)

run through a 2× decimation (line and sample) before compression. A 4:2:0[4] coding was used.

Figure 8.12 shows the effects of MPEG-2 compression on video captured with an 8-mm video camera, digitized, and displayed on a digital monitor. The original uncompressed data is plotted assuming an effective bandwidth of 30 Mbps. These results cannot be extrapolated to original digital formats because of differences in high-frequency content.

This same study also provided some initial data on the effects of compression algorithms. The data shown in Figure 8.12 was compressed using a software CODEC.[5] Figure 8.13 shows a comparison of data from the software CODEC and two hardware CODECs. The 6-Mbps hardware CODEC was an MPEG-2 implementation; the 1.54-Mbps hardware was an MPEG-1 implementation. Ratings were significantly higher for the hardware CODEC at 1.54 Mbps. Again, MPEG-2 is optimized at rates of 3–10 Mbps.

A second study [10] compared different analog video formats including 8 mm, Hi-8, and analog Betacam. Hi-8 video captures color (chroma)

---

4. The expression 4:2:0 denotes the number of luminance and chrominance blocks within a macroblock where a block is 8 lines and 8 pixels/line. The 4:2:2 format has four luminance blocks, two blue-luminance (B-Y) chrominance blocks, and two red-luminance (R-Y) chrominance blocks within a macroblock. The 4:2:0 format has four luminance, one B-Y, and one R-Y.

5. A CODEC is a compression coder/decoder algorithm.

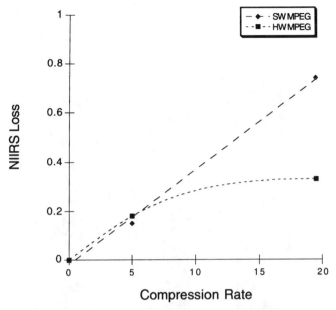

Figure 8.13 Hardware versus software CODEC results. (Data from [2].)

data in a separate channel, which avoids the encoding process required for composite video (video in which the chroma data is carried on sidebands above and below the subcarrier frequency used for the luminance data). Analog Betacam has an estimated resolution of 500 lines by 700 pixels and provides three-channel color (Y, B-Y, R-Y). The data was compressed to rates of 6 and 1.54 Mbps using the same procedure as used in the first study.

The study showed no significant differences in NIIRS ratings of the sample of uncompressed video clips (total of 22). Figure 8.14 shows the effects of the compression on the three formats. There is no significant difference among the three formats.

8.2 Enhancement Processing

The previous three sections have discussed image processing algorithms that generally degrade the interpretability of imagery. In this section, we discuss a class of algorithms or processes that improve the interpretability of imagery. These include spatial filtering (sharpening and noise reduction) and pixel intensity transforms (contrast enhancement processing). Collectively, these processes are called enhancements. The effect of enhancement processing is to make interpretation easier, not to actually add information. For example,

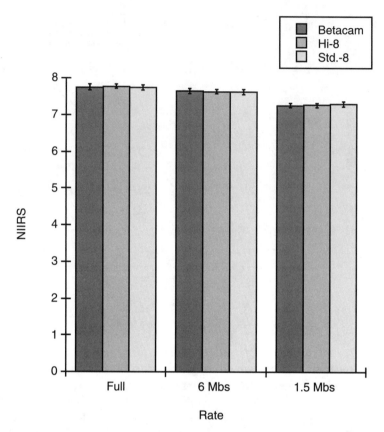

Figure 8.14   Effect of input format on compression loss. (Data from [10].)

dynamic range adjustment (DRA) stretches the image dynamic range of interest over the display range, thus making it easier to detect small contrast differences.

8.2.1   Spatial Filtering

Spatial filtering is also called modulation transfer function compensation or correction (MTFC). As an edge-sharpening device, MTFC is typically a convolver in the spatial domain, which boosts the edge as shown in Figure 8.15. A quadrant symmetric kernel is employed such as is shown in Figure 8.15. The kernel is convolved with the unsharpened image to produce a sharpened image. A 5×5 kernel is the recommended minimum size [15]. Larger sizes can be used at the expense of increased processing time. MTFC can also be implemented by multiplication in the spatial frequency domain.

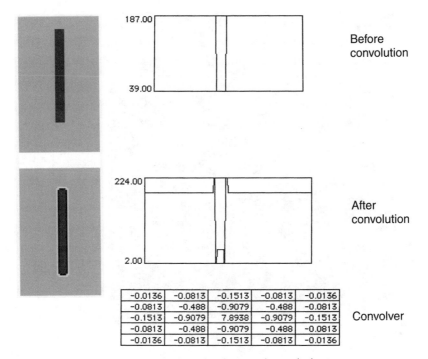

Figure 8.15 Effect of 5 × 5 sharpening filter showing convolver and edge traces.

Although the primary use of MTFC is to sharpen, it can also be used to reduce the effects of noise. This is accomplished by smoothing as opposed to sharpening. Conversely, when sharpening is applied, noise is also sharpened and perceptually increased.

Figures 8.16 and 8.18 provide examples of the effects of MTFC. In the first case, sharpening has been applied to sharpen the image. Note that although blur is reduced, the pavement shows increased noise. Figure 8.17 shows a trace across the same portion of both images. Note the increase in edge contrast as well as the increase in noise for the sharpened image.

In Figure 8.18, a smoothing filter has been applied to a noisy image. Note the vehicle at A. Figure 8.19 shows a trace through the same portion of each scene. The effect of the blur function is evident.

In the frequency domain, wavelet processing has been demonstrated as a SAR clutter reduction technique [16]. A noisy signal is decomposed into its wavelet representation, and then coefficients finer than some defined resolution level are set to 0.

The effect of MTFC on interpretability is difficult to quantify because it depends on the sensor characteristics and the interpretation task. In the

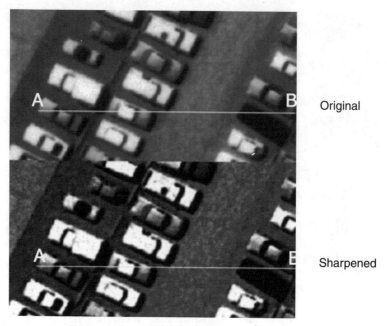

Figure 8.16   Effect of sharpening filter.

examples shown in Figures 8.16 and 8.18, the general appearance of the image was improved and some objects in the scene appeared easier to interpret. The same information, however, was in both versions of the scene. In most cases, the effect is likely to be on the order of 0.1 NIIRS or less. Under time-

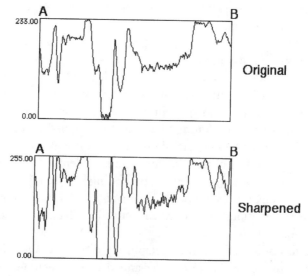

Figure 8.17   Trace across line A-B in Figure 8.16.

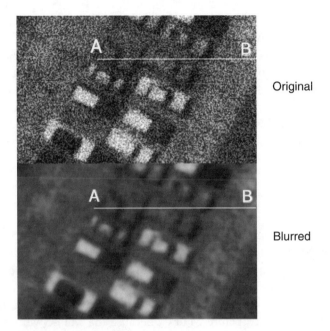

Figure 8.18 Effect of blur filter on noise.

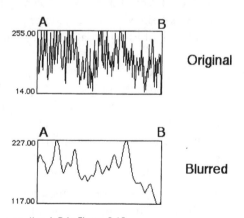

Figure 8.19 Trace across line A-B in Figure 8.18.

constrained tasks (e.g., recognition at range) the effect can be substantially greater, but again, it is system-, scene-, and task-dependent.

8.2.2 Pixel Intensity Transforms

Pixel intensity transforms modify the digital value of pixels, generally for the purpose of enhancing contrast. This process is typically performed in three steps. The first is to adjust the pixel intensity values from the sensor to the region of interest on the display. Most current displays are limited to 8 bits (256 levels); sensors (e.g., RADARSAT) can capture as many as 16 bits. In addition, portions of the sensor dynamic range may not be of as much interest as other portions. For an EO sensor, lower count values are dominated by haze. Elimination of these values improves image contrast. The form of DRA used for the Image Data and Exploitation (IDEX) system is a combined subtraction and multiplication process [15]. A scaled intensity value is defined by

$$I_{New} = M \times (I_{Old} - I_o) \qquad (8.3)$$

where I_{New} is the transformed pixel intensity value, M is a multiplier, I_{Old} is the original pixel intensity value, and I_o is the data base value. This data base value is a pixel intensity value associated with a particular type of imagery. The value of M and I_o can be based on the image histogram or can be system-dependent. If the values are based on the image histogram, the histogram must be defined before MTFC, because the MTFC will modify the histogram.

The effects of DRA on a typical image are shown in Figure 8.20. The effect on the histogram is also shown. Because the original histogram contained less than 256 levels, the histogram after DRA shows the effects of quantization.

The second step in the pixel intensity transform process involves selective enhancement of specific portions of the scene dynamic range. This is sometimes called tonal transfer compensation (TTC). This is a nonlinear transform that expands some portions of the dynamic range at the expense of others. Figure 8.21 shows an example with radar imagery. The nonlinear transform (shown as an inset) brings out detail on the vehicles. The effect on the image histogram is also shown in Figure 8.22. Such transforms can selectively enhance specific portions of the dynamic range.

The final step in the pixel intensity transform process entails what is called perceptual linearization. The eye does not respond to luminance changes in a linear fashion. At low luminance levels, a larger luminance difference (higher value of Cm) is required for detection than at high luminance levels. Perceptual linearization involves a nonlinear transform to the pixel intensities after DRA and TTC have been applied so that all gray-level difference be equally discriminable. This also has the effect of making different displays look the same, given the same input image. Figure 8.23 shows the transform developed for medical imagery based on a model of the human visual system [17]. The model was validated by comparison to previous sets

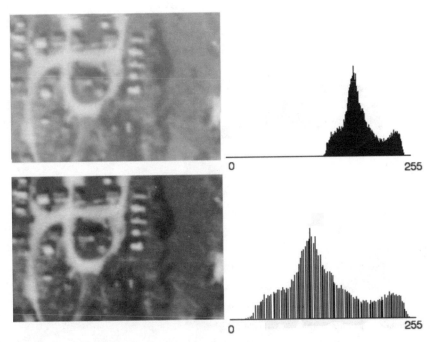

Figure 8.20 Effects of DRA.

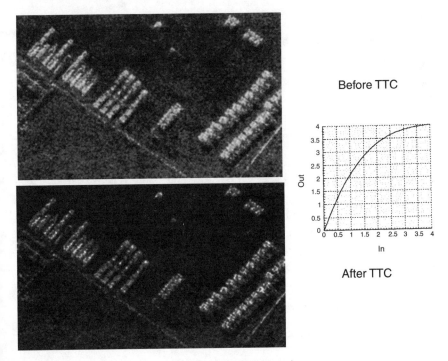

Before TTC

After TTC

Figure 8.21 Effect of TTC. (Image courtesy of TESAR Program.)

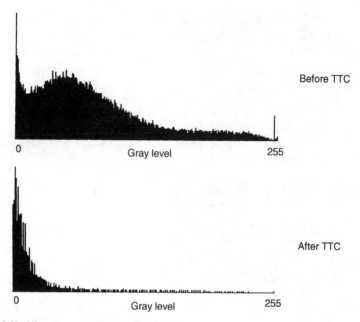

Figure 8.22 Histograms of images in Figure 8.21.

of empirical data. The equation was developed as a standard for Digital Imaging and Communications in Medicine (DICOM) by the National Electrical Manufacturers Association (NEMA) and the American College of Radiology. It is called the NEMA/DICOM standard and is defined as

$$\log_{10} L_i = \frac{a + c \ \ln(I) + e \ (\ln(I))^2 + g \ (\ln(I))^3 + k \ (\ln(I))^4}{1 + b \ \ln(I) + d \ (\ln(I))^2 + f \ (\ln(I))^3 + h \ (\ln(I))^4 + m \ (\ln(I))^5} \quad (8.4)$$

where $a = -1.3011877$, $b = -0.025840191$, $c = 0.080242636$, $d = -0.10320229$, $e = 0.13646699$, $f = 0.02874562$, $g = -0.025468404$, $h = -0.0031978977$, $k = 0.00012992634$, and $m = 0.00133635334$. I are digital count values ranging from 1 to 1023, and L is measured in cd/m^2.

A similar form is used for the IDEX system. The IDEX equation was developed on the basis of an empirical study and is defined as

$$L = 0.1 + 4.060264x - 6.226862 \ x^2 + 48.145864 \ x^3$$

$$- 60.928632 x^4 + 49.848766 \ x^5 \quad (8.5)$$

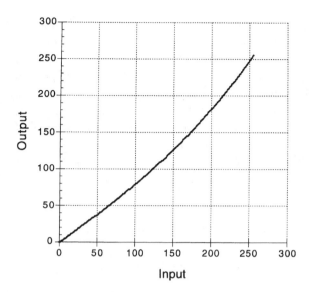

Figure 8.23 NEMA/DICOM calibration function.

where L = output luminance and x = normalized command level where $0 \leq x \leq 1.0$ and L varies from 0.1 to 35 fL. No difference in the two forms has been seen for visible imagery; the NEMA/DICOM form was shown to be superior for radar imagery [18].

The effect of pixel intensity transforms depends on the characteristics of the input image and the interpretation task involved. Probably the most significant effect is that of limiting dynamic range. From a study of softcopy interpretability under conditions of varying dynamic range, a loss of up to 0.8 NIIRS can be inferred [19]. The effect of perceptual linearization is much smaller, on the order of 0.1 NIIRS or less. In another example, two pixel intensity transforms were applied to IR imagery. One (histogram projection, a nonlinear DRA) improved P_D values by ~4%, the other (histogram equalization) decreased P_D by ~20% [20].

The pixel intensity transforms discussed were developed for monochrome imagery. The same transforms are applicable to color imagery (MSI, HSI) for general viewing. The NEMA/DICOM perceptual linearization function has also been shown to apply to MSI display [19].

8.2.3 Image Fusion

Multispectral systems often trade resolution for spectral diversity. IKONOS, for example, has four 4-m GSD spectral bands plus a panchromatic band with

1-m GSD. A technique called band sharpening can be used to improve the apparent GSD of the spectral images. An issue is the degree to which the sharpening or fusion preserves the spectral integrity of the data. The benefit of the sharpening is a function of the similarity of the pan and MS images in terms of imaging geometry and atmospheric conditions. The extreme case is one in which the technique is demonstrated by creating a pan image from a set of MS images and fusing the pan image with coarser GSD versions of the MS image. In such cases, image analysts estimate that 94% of the pan resolution (GSD) is preserved in the fused product [21]. For example, if the GSD of the pan is 1m and the MS is 5-m, the effective GSD of the fused product is 1.24m:

$$\text{Effective GSD} = \text{GSD}_{MSI} - 0.94(\text{GSD}_{MSI} - \text{GSD}_{Pan}) \qquad (8.6)$$

In NIIRS space (MS IIRS), the effect was defined as

$$\text{Delta-MS IIRS} = 3 \log(\text{GSD}_{MSI}/\text{GSD}_{Pan}) - 0.3 \qquad (8.7)$$

With the 5-m MS and 1-m pan, the predicted improvement in MS IIRS (relative to the 5-m MS IIRS) is 1.8.

A worst-case situation exists when the pan and MS images were obtained at different times by different sensors. SPIN-2 images (nominal 2-m GSD) were used to sharpen LANDSAT MS images at 28-m GSD. The MS IIRS improvement was 0.9 MS IIRS as opposed to the 3.1 predicted by (8.3) [22].

Fusion can also be applied to images collected in different IR bandwidths (SWIR, MWIR, LWIR). One such study showed a 3–10 times improvement in detection performance over use of a single IR band [23]. Finally, fusion can be applied to images collected in radically different portions of the spectrum. Here, the problem can be complicated by different imaging geometries (e.g., radar and visible or IR).

8.3 Summary

The effects of bandwidth compression range from negligible to severe depending on the nature of the original imagery, the degree of compression, and the method of compression. For still imagery, wavelet compression has been shown to be superior to DCT forms. A loss of less than 0.2 NIIRS can be maintained with an 8:1 wavelet (JPEG 2000) compression. For video imagery, MPEG-2 is the preferred compression for rates of 3–10 Mbps; MPEG-1 may be preferred for a 1.54-Mbps rate, although MPEG-4 is being designed for lower rates.

Spatial filtering is used to improve edge sharpness and to reduce the effects of noise. The effect is generally small, on the order of 0.1 NIIRS. Pixel intensity transforms are used to optimally allocate the dynamic range of the original image to the output display. In extreme cases, reducing the output dynamic range can result in a loss of greater than 0.5 NIIRS. Assuming a reasonable attempt at optimization, however, the effect of variations is on the order of 0.1 NIIRS or less.

References

[1] Leachtenauer, J. C., K. Daniel, and T. P. Vogl, "Digitizing Corona Imagery: Quality vs. Cost," in R. A. McDonald (ed.), *Corona Between the Sun and the Earth: The First NRO Reconnaissance Eye in Space*, Bethesda, MD: American Society for Photogrammetry and Remote Sensing, 1997.

[2] Leachtenauer, J., M. Richardson, and P. Garvin, "Video Data Compression using MPEG-2 and Frame Decimation," *Proceedings of SPIE Visual Information Processing VIII*, Vol. 3716, Orlando, Florida, 1999, pp. 42–52.

[3] Linne von Berg, D. C., and M. R. Krueer, "Image Compression for Airborne Reconnaissance," *Proceedings, SPIE Conference on Airborne Reconnaissance XXII*, Vol. 3431, San Diego, California, 1998, pp. 2–13.

[4] Franti, P., T. Kaukoranta, and O. Nevalainen, "Blockwise Distortion Measure in Image Compression," *Very High Resolution and Quality Imaging*, SPIE, Vol. 2663, pp. 78–87.

[5] Kaukoranta, T., P. Franti, and O. Nevalainen, "Empirical Study on Subjective Evaluation of Compressed Images," *Very High Resolution and Quality Imaging*, SPIE, Vol. 2663, pp. 88–99.

[6] Wallace, G. K., "The JPEG Still Picture Compression Standard," *Communications of the ACM*, April 1991, pp. 30–44.

[7] Briggs, S. J., *Manual: Digital Test Target BTP#4*, D180-25066-1, Seattle, WA: Boeing Aerospace Company, 1979.

[8] Rajan, S. D., A. T. Chien., and B. V. Brower, "Advanced Commercial Compression that Will Enable USIGS to Meet the Requirements of TPED and JV2010," paper presented at the 1999 IS&R conference, Washington DC, 1999.

[9] Novak, K., and F. S. Shahin, "A Comparison of Two Image Compression Techniques for Softcopy Photogrammetry," *Photogrammetric Engineering and Remote Sensing*, Vol. 62, No. 6, 1996, pp. 695–701.

[10]   Leachtenauer, J., M. Richardson, and P. Garvin, *Video Camera Comparison Study*, Merrifield, VA: National Imagery and Mapping Agency, 1997.

[11]   Saghri, J. A., A. G. Tescher, and A. Boujarwah, "Spectral-signature-preserving Compression of Multispectral Data," *Optical Engineering*, Vol. 38, No. 12, 1999, pp. 2081–2088.

[12]   Coleman, G., *MSI Data Compression Design Trade Study*, Washington, DC: National Exploitation Laboratory, 1995.

[13]   Strachan, D. "Video Compression," *SMPTE Journal*, February 1996, pp. 68–73.

[14]   DoD/IC/USIGS Video Working Group, *Video Imagery Standards Profile*, Version 1.5, October 20, 1999.

[15]   Central Imagery Office, *USIS Standards and Guidelines*, Appendix IV-Image Quality Guidelines, Vienna, VA: Central Imagery Office, 1995.

[16]   Horgan, G., "Wavelets for SAR Image Smoothing," *Photogrammetric Engineering and Remote Sensing*, Vol. 64, No. 12, 1998, pp. 1171–1177.

[17]   NEMA/DICOM, *Grayscale Standard Display Function*, PS 3.14-1998, National Electrical Manufacturers Association, January 1998.

[18]   Leachtenauer, J. C., G. Garney, and A. Biache, "Effects of Monitor Calibration on Imagery Interpretability," *Proceedings of the PICS Conference*, Portland, Oregon, 2000, pp. 124–129.

[19]   Leachtenauer, J. C., and N. L. Salvaggio, "Color Monitor Calibration for Multispectral Displays," *Digest of Technical Papers, Society for Information Display*, Vol XXVIII, Boston, Massachusetts, 1997, pp. 1037–1040.

[20]   Aviram, G., "Evaluating the Effect of Infrared Image Enhancement on Human Target Detection Performance and Image Quality Judgement," *Optical Engineering*, Vol. 38, No. 8, 1999, pp. 1433–1440.

[21]   Vrabel, J., "Multispectral Imagery Advanced Band Sharpening Study," *Photogrammetric Engineering and Remote Sensing*, Vol. 66, No. 1, 1998, pp. 73–79.

[22]   Civil and Commercial Applications Project, *SPIN-2 Evaluation*, Bethesda, MD: National Imagery and Mapping Agency, 2000.

[23]   Scribner, D., et al., "Infrared Color Vision: An Approach to Sensor Fusion," *Optics and Photonics News*, August 1998, pp. 27–32.

Selected Bibliography

Fibush, D. K., R. Elind, and K. Ainsworth, *A Guide to Digital Television Systems and Measurements*, Portland, OR: Tektronix, Inc., 1994.

Inglis, A. F., and A. C. Luther, *Video Engineering*, New York, NY: McGraw Hill, 1996.

International Telecommunication Union, Recommendation ITU-RBT.601, *Encoding Parameters of Digital Television for Studios*, 1994.

International Telecommunication Union, Recommendation ITU-RBT.624, *Characteristics of Television Systems*, 1986.

Society of Motion Picture and Television Engineers, SMPTE 170M, *Composite Analog Video Signal, NTSC for Studio Applications*, White Plains, NY, 1994.

Society of Motion Picture and Television Engineers, SMPTE 259M, *10 Bit 4:2:2 Composite and 4fsc Composite Digital Signals–Serial Digital Interface*, White Plains, NY, 1995.

Welstead, S., *Fractal and Wavelet Image Compression Techniques*, Bellingham, WA: SPIE, 1999.

9

Display and Observer Considerations

In this chapter, we reach the end of the image chain at the display and observer. We consider the effect of display variables related to hardware. We also consider the effects of the display environment. Viewing imagery on a cathode ray tube (CRT) in normal office lighting can reduce interpretability by up to half a NIIRS. Acoustic noise and image motion can also affect performance, although little data exist to quantitatively describe their effects.

The observer is first considered from the viewpoint of selection. What attributes are associated with a proficient observer? What are the effects of training? We show data on typical observer variability. Finally, models of the visual system are reviewed.

9.1 Displays

Displays are characterized as hardcopy (HC) or softcopy (SC). Hardcopy display output includes film and paper prints. Softcopy includes CRTs and a variety of backlit digital displays such as liquid crystal displays. For many years, HC was the standard method of displaying S&R imagery, and some models still assume a HC display (e.g., the GIQE [1]). Currently, however, SC is becoming the dominant media.

SC has two potential advantages over HC. First, the dynamic range can be greater. For example, a film density range of 0.3 to 2.0 density units is a

17-dB range.[1] Monochrome SC monitors can show ranges of 30 dB or more. Second, the relationship between digital input values and output luminance can be modified. In one of the first comparisons of HC and SC, the softcopy was preferred over the hardcopy [2]. This was an unexpected finding, given that the softcopy images were produced by digitizing the hardcopy. The difference increased as the judged quality of the imagery decreased.

Approximately 20 years later, a similar finding was shown with digitized Corona imagery [3]. When the Corona imagery was digitized with a 4-μm spot size frame-grabber, the NIIRS of the softcopy was slightly higher than the hardcopy. The difference was on the order of 0.1 NIIRS. Similar results were seen in a second study [4]. In a third study, the average difference for visible imagery was 0.3 NIIRS; the difference for radar was not statistically significant [5]. Results are shown in Figure 9.1. On an individual image basis, results can vary substantially. In the Corona study, differences ranged from −0.3 to + 0.8 NIIRS. In the third study differences ranged from −0.25 to +0.78 [5].

Substantial variability exists among softcopy displays. The issue becomes confusing with the number of different measures used to characterize displays. The National Information Display Laboratory (NIDL), for example, uses 13 different measures of monitor quality [6]. Each one of these measures can have multiple values as a function of screen position and commanded luminance. A flat panel display measurements standard lists some 42 photometric and colormetric measures recommended for display characterization [7]. The measures used to characterize displays can be classified as luminance measures, resolution measures, and spatial/temporal measures. In the physical domain, such measures are defined by photometric, distance, and size measurement, and in some cases as a function of time. Perceptual targets are also used to assess monitor performance. One of the most useful is the Briggs target [8].

9.1.1 Briggs Targets

Briggs targets are used to assess contrast as a function of resolution. Results are expressed in terms of a Briggs score. Briggs scores are defined by reading the smallest resolvable patterns in a Briggs target [8]. The eight-level C-7 target set is shown in Figure 9.2. The eight levels or targets are equally spaced across the command level continuum, in this case, 8 bits. The light and dark

---

1. Density is defined in log units because density $= \log(1/T)$, where T is the ratio of incident to transmitted light. Density is thus the log of a ratio. A density difference of 1.7 is 17 dB ($10 \log_{10}(\Delta D)$).

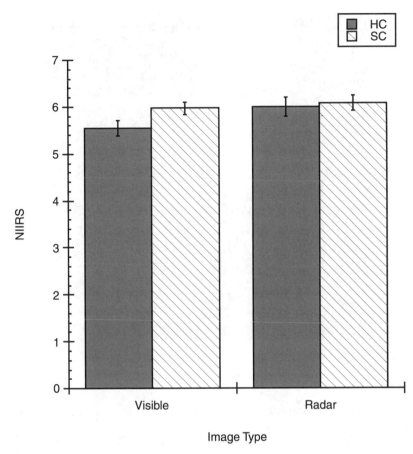

Figure 9.1 Hardcopy versus softcopy NIIRS ratings. (Data from [5].)

squares in the targets differ by 1, 3, or 7 digital counts and are called respectively the C-1, C-3, and C-7 targets. The backgrounds are set to 0.7 times the average count of the dark and light squares in each board. The number of checkerboards per target and pixels per square vary as shown in Table 9.1 and Figure 9.3. Table 9.1 also shows the Briggs score for each board. The observer's task is to define the smallest board where the light and dark squares can be resolved. The observer is then asked to rate the appearance of the target as shown in Figure 9.4. The 1–5 appearance rating is subtracted from the base score listed in Table 9.1 to provide the final score. Observers are allowed to move with respect to the display and, in the original instructions, to use optical magnification. In all of the studies cited here, magnification was not allowed.

The number of squares per board in the Briggs target was varied to provide a better continuum than one that simply varied the number of pixels

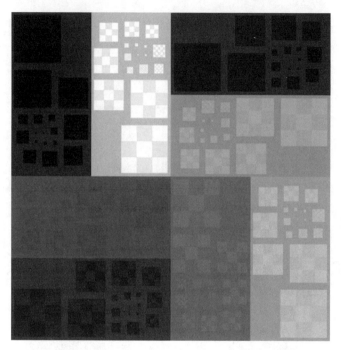

Figure 9.2 Briggs target set.

per square. The relationship was developed empirically using a 1970s-era 50 pixels per inch (ppi) monitor [9]. For this reason, the target seems to work best with monitors of the same pixel density; comparing monitors of different pixel densities is more difficult.

The Briggs score is valuable in that it can be provided by observers with little training and it correlates highly with NIIRS ratings. It is also often more sensitive than NIIRS to display differences.

9.1.2 CRT Displays

The number of measures that significantly affect performance on CRTs is quite small. Table 9.2 lists the measures that have been shown to significantly impact CRT interpretability as measured by NIIRS ratings and Briggs scores.

Dynamic range (DR) is defined as

$$DR = 10 \, Log_{10}(L_{max}/L_{min}) \tag{9.1}$$

where L_{max} is the maximum output luminance and L_{min} is the minimum.

Table 9.1
Briggs Target Characteristics

Target #	Base score	Pixels/square	Squares
B-10	10	25	5
B-15	15	20	3
B-20	20	16	3
B-25	25	13	3
B-30	30	10	3
B-35	35	8	3
B-40	40	7	3
B-45	45	4	5
B-50	50	3	5
B-55	55	2	7
B-60	60	2	5
B-65	65	1	11
B-70	70	1	7
B-75	75	1	5
B-80	80	1	4
B-85	85	1	3
B-90	90	1	2

Cm is defined as

$$Cm = \frac{L_B - L_D}{L_B + L_D} \qquad (9.2)$$

where Cm is contrast modulation, L_B is the brightest of two luminance values, and L_D is the darker of two values. Pixel density[2] is simply the number of pixels per inch and bit depth (BD) is

$$BD = \log_2 CL \qquad (9.3)$$

where CL is the number of command levels.

---

2. Pixel density is defined as the number of active or commandable display pixels per unit dimension (e.g., 100 pixels/inch (ppi)).

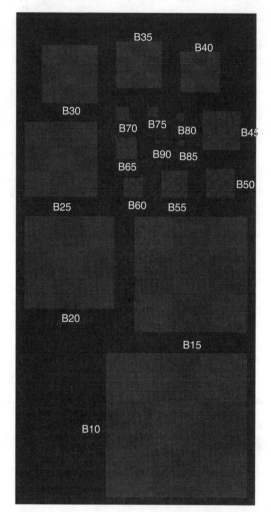

Figure 9.3 Briggs target.

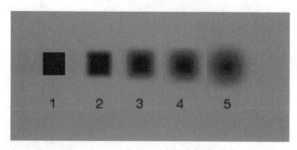

Figure 9.4 Briggs target quality rating.

Table 9.2
Critical CRT Measures

Type	Measure	Threshold value	Desired value
Luminance	Dynamic range	22 dB	25.4 dB
	Lmax	16 fL	D 35 fL
Resolution	Cm	35%	D 50%
	Pixel density	72 ppi	< 140 ppi
	Bit depth	8	8

The values shown in Table 9.2 were derived from a series of studies that compared different monitors and monitor calibrations. As will be discussed shortly, the values include the effects of ambient light. With current CRT technology, ambient light must be very low in order to achieve the desired levels.

Figure 9.5 shows the effects of dynamic range for IR and visible imagery NIIRS [4]. Clearly a significant loss has occurred at 8.4 dB. For the visible, the loss is significant at 20.6 dB. For the IR, the threshold is below 20 dB. Figure 9.6 shows Briggs data (C-7 target), where the threshold is again below 20 dB.

Figure 9.7 [10] shows the results of a similar study with MSI. A second-order polynomial has been fit to the data. In this case, the threshold appears to be on the order of 20 dB.

Moving in the opposite direction, Figure 9.8 shows no significant improvement when the dynamic range is increased beyond the recommended 25.4 dB [5]. For the display at 38.5 dB, the maximum luminance value was set at 107 fL and some analysts complained about the high brightness of the monitor.

The recommended values for dynamic range and L_{max} represent somewhat of a tradeoff between comfort and good vision. The eye works in three zones, the photopic, mesopic, and scotopic, as shown in Figure 9.9 [11]. In order to work solely in the photopic region and still maintain the desired dynamic range, it is necessary to use very high brightness levels (350 fL). Not only are such values beyond the range of most CRTs, but they are also uncomfortable for long periods of viewing. It is therefore necessary to extend the display dynamic range into the mesopic region, which brings the rods of the eye into play; they are more sensitive but provide less resolution. The scotopic region is avoided.

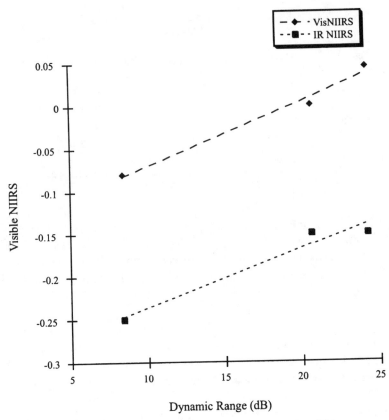

Figure 9.5 Effect of display dynamic range on NIIRS ratings. (Data from [4].)

Using the NEMA/DICOM model (8.4), the number of JNDs over a defined luminance range can be calculated [12]. For the recommended range of 0.1–35 fL, 453 JNDs are predicted. This, of course, is more than an 8-bit display can portray. Referring again to Figures 9.5 and 9.6, the NEMA/DICOM model indicates 387 JNDs at a dynamic range of 20.6 dB and only 217 at 8.6 dB. The NEMA/DICOM model also suggests why there is no benefit from increasing dynamic range when the number of JNDs already exceeds the capability of the display. Finally, note that 363 JNDs result when the recommended minimum value of dynamic range and Lmax are used.

Contrast modulation defines the ability to resolve fine detail. The Cm threshold varies as a function of spatial frequency, as shown in Figure 9.10 [13]. The referenced data was replotted to show spatial frequency in terms of pixel density. Points above the line have the necessary contrast to be detected; those below do not. A 100-ppi monitor at 2× magnification does not

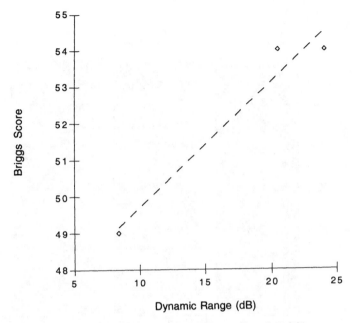

Figure 9.6 Effect of display dynamic range on Briggs scores. (Data from [4].)

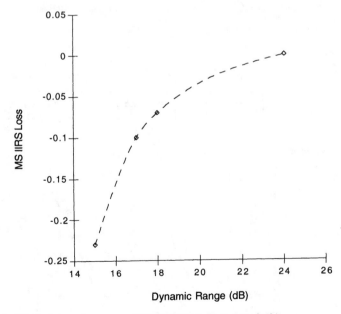

Figure 9.7 Effect of dynamic range on MS IIRS ratings. (Data from [10].)

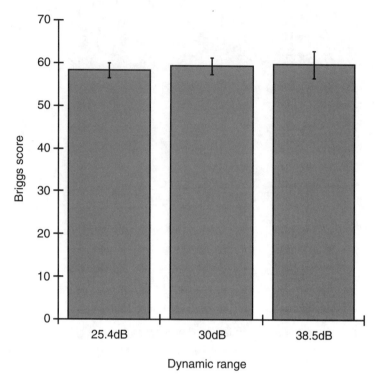

Figure 9.8 Effect of expanded dynamic range on Briggs scores. (Data from [5].)

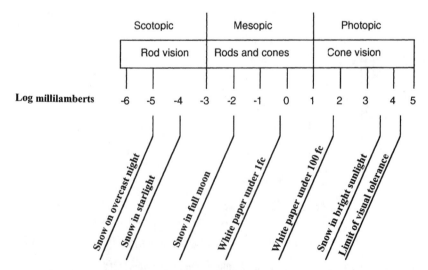

Figure 9.9 Visual illuminance zones.

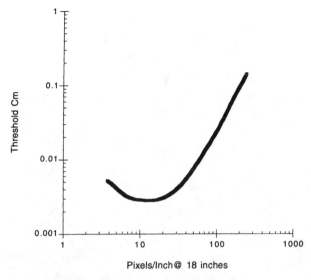

Figure 9.10 Visual sensitivity function.

allow detection of Cm values smaller than 0.01. The Cm threshold is lowest at a pixel density of ~17 ppi. Most viewers, however, do not want to view imagery at that pixel density. Figure 9.11 shows an image enlarged (with bilinear interpolation) to simulate pixel densities of 72, 36, and 18 ppi.

Contrast modulation performance is measured using what is called an aperture grill (Figure 9.12) commanded to full on and full off. Typically, line width is varied from one to three pixels and Cm is measured at various points on the monitor. For CRTs, Cm is typically highest in the center of the monitor where focus is best and it falls off in the corners. Cm values typically range from 15–60% between and within different color monitors [14]. Values of 25% are considered adequate for viewing imagery and 50% for text (as measured using a 100% Cm target) [14]. Color monitors have inherently lower Cm values than monochrome monitors because they use shadow masks and aperture grills to produce color. A single pixel is made up of three color dots or lines. Cm is typically significantly lower in one dimension than the other for a color monitor.

Contrast modulation can be effectively improved with magnification. Figure 9.13 shows Cm data for four monitors at three grill sizes. A 2× magnification of the 1/1 grill will achieve essentially the same results as the 2/2 grill and thus a substantial improvement in Cm. Note that the results are not uniform; there is not a consistent relationship between grill size and Cm.

A comparison of NIIRS ratings on a color (shadow mask) and monochrome monitor, both with 100-ppi addressability, is shown in Figure 9.14

72 ppi

36 ppi

18 ppi

Figure 9.11 Effect of enlargement.

Figure 9.12 Aperture grill pattern used to measure Cm.

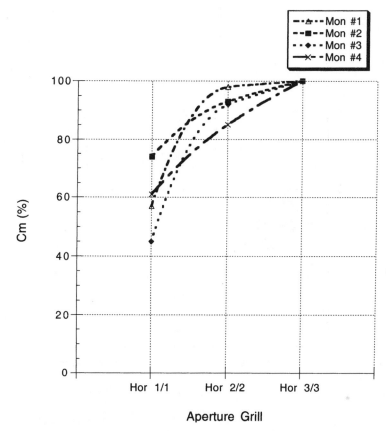

Figure 9.13 Effect of magnification on Cm.

[4]. At 1× magnification, results are better on the monochrome monitor. At 2×, there is no significant difference between the two.

A comparison of two monochrome monitors is shown in Figure 9.15. Because of design differences, the two monitors differed in terms of Cm performance. Monitor 1 had Cm values ranging from 35–62%; monitor 2 values ranged from 50–80%. There were no significant differences in NIIRS ratings or Briggs scores [15].

Pixel density recommendations are based on the results of two studies. Figure 9.16 shows the effects of pixel density on visible imagery delta-NIIRS ratings where pixel density was varied by magnification [4]. Performance at 50 ppi (2× magnification using bilinear interpolation) exceeds that at 100 ppi. In a second study, four levels of pixel density were evaluated [16]. Figure 9.17 shows the effect of pixel density on Briggs scores; Figure 9.18 shows the effect on NIIRS ratings. Pixel density is shown in log units.

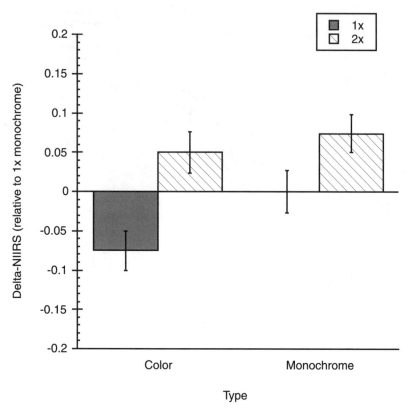

Figure 9.14 Comparison of monochrome and color monitors at two magnifications. (Data from [4].)

The nature of Briggs targets is that they are insensitive to blocking effects with magnification. Consequently, Briggs scores increase with magnification so long as pixel replication is used (as it was in this study). NIIRS ratings (and the imagery on which they are made) are sensitive to blocking effects—bilinear or some other type of interpolation is applied to imagery when it is magnified to avoid blocking. Figure 9.18 shows peak NIIRS performance in the region of 70–80 ppi. A 4× magnification of a 170-ppi monitor has apparently degraded interpretability. At higher pixel densities, the image detail cannot be resolved, and thus, interpretability is degraded relative to some lower pixel density.

Bit depth is defined by the processing of data as the last digital step before the display. In many cases, imagery can be processed at larger bit depths and is reduced to 8 bits only as the last step before display. The current bit depth standard is 8 bits. Some software applications reduce bit depth by up to some portion of 1 bit by reserving a portion of the dynamic range for color.

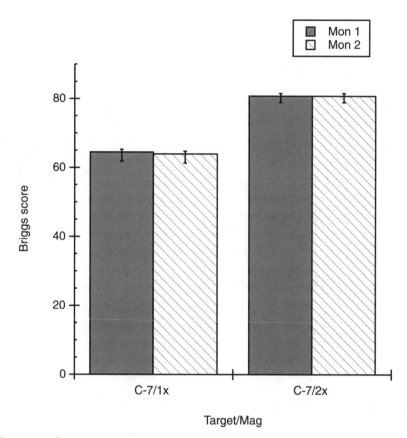

Figure 9.15 Comparison of two monochrome monitors. (Data from [15].)

No studies have been performed to show a significant interpretability loss for this reduction, but a loss is theoretically possible. Some types of imagery have as many as 16 bits per channel (e.g., RADARSAT), and it can be hypothesized that larger bit depths would be advantageous. To date, however, no studies have been performed to demonstrate such an advantage.

In addition to the critical CRT measures given in Table 9.2, there are desirable measures that should be met for a softcopy display. These measures have not been shown to directly affect performance, but fatigue and annoyance can occur if they are not met. These requirements are shown in Table 9.3.

Luminance is not uniform across the face of a CRT. Fortunately, there is seldom a need to compare, in absolute terms, the luminance of widely scattered points or areas. The human visual system is not capable of high accuracy in performing such comparisons [10]. The luminance uniformity issue

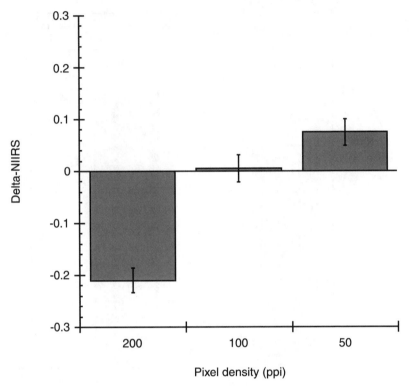

Figure 9.16   Effects of pixel density on delta-NIIRS. (Data from [4].)

is thus more one of maintaining dynamic range across the face of the monitor. Halation is a measure of internal reflections. Again, the issue is one of reducing contrast. Color temperature (white point) values at the two levels shown have been shown to have no significant effect on the interpretability of multispectral imagery [10]. Step response is the response of the monitor as the electron beam crosses an edge. No overshoot or undershoot should be present. A monitor that has difficulty in the transition may exhibit banding or ringing after the transition. Straightness is defined in terms of non-linearity of a horizontal or vertical straight line across the face of a monitor. Linearity is the relationship between the actual and commanded position of a pixel on the display. The monitor size recommendation is based on a desire to have at least a 1024 × 1024 pixel image displayed on the monitor at no greater than 100 ppi and a desire to avoid the need for head motion to view all of the screen.

    Jitter, swim, and drift are movements of individual pixels over short time durations. These movements can be very annoying if they are notice-

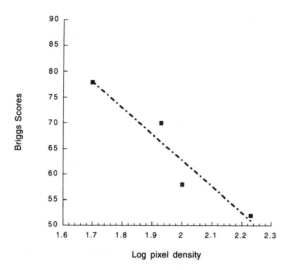

Figure 9.17 Effects of pixel density on C-7 Briggs scores. (Data from [16].)

able. For most applications, the issue is one of annoyance rather than de-graded interpretability. Finally, refresh rate defines the rate at which the image on a screen is updated. If refresh rate is too low, flicker occurs. A rate of 60 Hz is generally considered the minimum but the rate at which flicker[3] can occur varies as a function of image characteristics and ambient light [13].

9.1.3 Flat Panel Displays

Although CRTs are currently the dominant softcopy display, flat panel displays are becoming more prevalent. Flat panel displays refer to any one of several types of displays including liquid crystal (LCD), plasma, and electroluminescent displays.

In a flat panel display, the intensity controlling mechanism is embedded in the face of the display. The display consists of embedded rows and columns (picture elements or pixels) of a light-emitting or controlling layer [17]. The intensity (or color) of each pixel defines the image. Because there is no need to separate the controlling element and the varying intensity image, the display can be flat.

CRTs have historically not been flat because of the method used to create the display. In a CRT, an electron beam created by heating a cathode

---

3. Flicker is the sensation of rapidly changing brightness on the face of a monitor. It is most apparent at high luminance levels, and some people have a lower threshold than others.

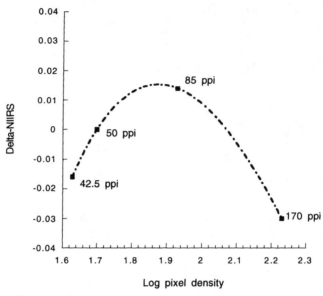

Figure 9.18 Effects of pixel density on delta-NIIRS ratings. (Data from [16].)

is focused and swept across the face of the monitor [17]. The electron beam excites a phosphor that emits light. The cathode and focusing mechanism is physically displaced from the phosphor coating on the face of the monitor. The current to the cathode controls the intensity of the displayed image. To facilitate focus, the face of the monitor typically exhibits some degree of curvature.

Table 9.3
Desirable CRT Measures

Type	Measure	Threshold value	Desired value
Luminance	Nonuniformity	≤ 35%	≤ 28%
	Halation	≤ 3.5%	≤ 2%
	Color temperature	≥ D65, ≤ D93	
	Step response	No ringing	
Resolution	Size	≥ 17.5 in, ≤ 24 in	
Spatial/temporal	Straightness	< 0.5%	
	Linearity	< 1%	
	Jitter/swim/drift	< 2 mils	
	Refresh rate	≥ 72 Hz	

The term "flat panel" is generally applied to non-CRT displays. Flat CRTs, however, have been under development for several years. Beam bending and matrix addressing techniques have been used to create flat CRTs, and so-called flat display CRTs are currently marketed. Matrix addressing consists of using a large-area cathode and switching grid to define the area of the screen illuminated. The field emissive display (FED) is another type of flat CRT display; it uses a field emitter cold cathode for each pixel.

Flat panel displays should meet the same requirements as CRT displays in terms of image quality. Flat panel displays have exhibited three problems relative to CRTs. The first is a lack of dynamic range. Early displays were limited to L_{max} values of less than 30 fL, and typical contrast ratios were on the order of 100:1 (20 dB) as opposed to the desired 350:1 or 25.4 dB. Recent active matrix liquid crystal displays (AMLCD) have largely overcome this problem. A second issue is the failure of individual pixels due to manufacturing defects. These failures are considered artifacts and, unless they are excessive, have little effect on interpretability. The manufacturing process has improved to the point where these artifacts seldom occur. The third issue is a decrease in contrast in off-axis viewing. Early laptop computers showed this problem—sufficient contrast was available only over a very small field of view. Current flat panel displays offer wider fields of view, although still not as wide as CRTs. There are two schools of thought with respect to this issue. One is that it is still a significant problem described by a reduction in dynamic range in off-axis viewing or, alternatively, a need to restrict the position of the head to maintain dynamic range. A second is that it can be beneficial in that it allows the observer to reduce very high contrast or brightness portions of the image by movement of the viewing point (personal communication, Mike Grote, National Imagery and Display Laboratory, February 1999). The issue has not been resolved.

Measurement requirements for flat panel displays have not yet been defined. At a minimum, flat panel displays should meet the same requirements as defined in Tables 9.2 and 9.3. In addition, there should be no significant number of manufacturing artifacts, and dynamic range should be maintained over some relatively large viewing angle. The number of allowable defects and the required viewing angle are not yet defined. Barring these issues, there is no reason to believe there will be any significant performance differences between flat panel and CRT displays.

9.1.4 Display Environment

Ambient light is a major source of image quality degradation. Ambient light reflects off the face of the monitor and adds to the luminance of the displayed image. The additive effect is constant regardless of the image luminance. A dynamic range of 0.1–35 fL becomes 0.9–35.9 with the addition of a 0.8-fL contribution from ambient light. This has two effects. First, the contrast between adjacent command levels is reduced in the lower portion of the command level range as shown in Figure 9.19. Second, the number of contrast JNDs is decreased because of the reduction in dynamic range. In the example given previously, roughly 20% of the JNDs available in the 0.1–35-fL range are lost with the addition of 0.8 fL from ambient light.

The effect of ambient light varies as a function of the sources and intensities of ambient light as well as the characteristics of the monitor (dynamic range, reflectance, and halation). An ambient light level of no greater than 1–2 fc is considered acceptable for CRTs, where the maximum luminance of color monitors is typically not much greater than 35 fL, and for monochrome monitors is not much greater than 150 fL. Normal office light levels of 20–40 fc are not considered acceptable. Flat panel displays can achieve higher brightness levels and can thus maintain a reasonable dynamic range (25 dB) in the presence of ambient light. Because of the high brightness levels required (>200 fL), however, fatigue and annoyance may be issues.

The decrease in dynamic range and loss of contrast with higher levels of ambient light degrade interpretability. A comparison of two ambient light

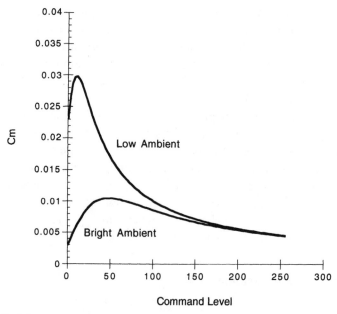

Figure 9.19   Effect of ambient light on Cm values.

levels (2 and 18 fc) showed a significant NIIRS loss for visible and radar imagery at the higher ambient level [5]. A sample of 10 radar and 10 visible images were displayed on a monitor calibrated to a dynamic range of 0.1–35 fL. In the 18-fc environment, the monitor luminance was increased by 0.8 fL, resulting in an actual dynamic range of 0.9–35.9 fL. Results are shown in Figure 9.20. The loss for individual images was as high as 0.6 NIIRS. The loss is greater for images with a preponderance of dark tones (water, shadows, and radar) and is greater as the ambient light contribution increases. A loss of 0.5 NIIRS for visible imagery is equivalent to a 40% decrease in resolution (defined in terms of GSD). This is in turn equivalent to a 40% decrease in the range required to achieve the same level of interpretability.

Image motion also can degrade interpretability. Two types of motion can occur. For some imaging systems, a so-called waterfall presentation is

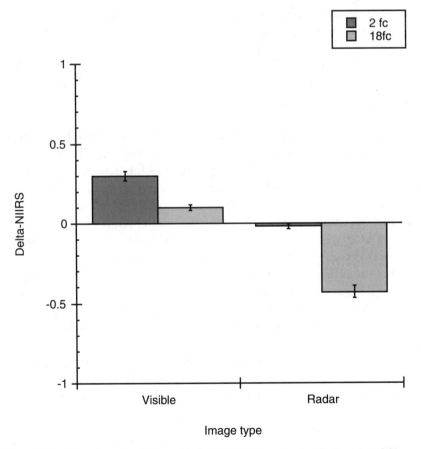

Figure 9.20 Effect of ambient light on delta-NIIRS ratings (relative to HC). (Data from [5].)

provided. The sensor moves continuously over the ground, and the imagery moves continuously from the top to the bottom of the display. The time that a given point is in the field of view is limited and, if the display rate is sufficiently fast, the display may be blurred resulting in an effective loss of resolution. The ability to resolve detail in such situations is termed dynamic visual acuity and varies across individuals. In a study requiring the detection of Landolt Cs,[4] the best observers were able to detect targets at a rate of 110 degrees/second, five times smaller than the worst group of observers [18]. Their dynamic visual acuity was thus said to be five times better.

The rate at which blur occurs, as measured by a decrease in the ability to resolve small detail, is on the order of 20–40 degrees/second [13]. This is equivalent to viewing a 21-inch display on which the scene changes every 1–2 seconds. If we assume a sensor with a resolution (GSD or IPR) of 1 ft , a 20-degree/second display rate equates to vehicle (sensor) velocity of 436 miles/ hour. The dominant effect on performance is more likely to be the limited time available. In performing search, the eye successively fixates on points in the image. The duration of these fixations is on the order of 0.2–0.3 seconds [19]. Ignoring interfixation time, in 1 second the eye can fixate 3–5 points. Unless targets are very large and conspicuous, a 1–2 second viewing time is not sufficient to search for and recognize targets on a 17-inch or larger display.

The effects of image motion can be demonstrated using the results of a study in which moving vehicle models were displayed on a TV display [20]. The observer's task was to distinguish between various vehicles (bus, large and small truck) and nonvehicle objects. Side views of the objects were presented and the background was not cluttered. Figure 9.21 shows results. Even with this very simple task, performance degrades as target motion increases from 10 to 20 degrees/second. In another study using moving radar imagery, target detection performance decreased by 15% when the display rate increased from 0.5 to 1.4 degrees/second [21]. This decrease was clearly an effect of the shorter time available for search and recognition. The effect of image motion at rates below the blur threshold is a function of image characteristics (scale, resolution, contrast), the task to be performed, and target conspicuity.

Motion can also occur due to vibration of the display or the observer and the display (as for example in a moving vehicle). As a rule of thumb, vibration amplitude should not exceed one-fourth the display resolution. For

4. A Landolt C is a ring with a gap equal in distance to the stroke width of the ring. The diameter of the ring is four times the gap dimension. In the typical study, the gap is shown at varying orientations and the observer's task is to identify the orientation of the gap.

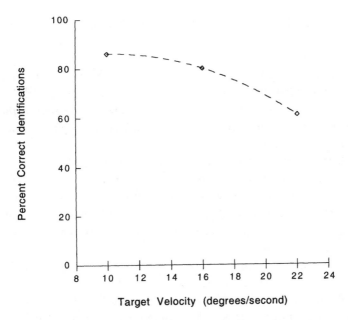

Figure 9.21 Effects of target velocity on vehicle identification. (Data from [20].)

a 100-ppi display, this would mean that vibration should not exceed 0.0025 inches [13]. A study of the ability to resolve gratings showed a performance degradation at amplitudes of 0.25 cycle and greater and also showed larger losses at higher frequencies (20 as opposed to 10 Hz). The effect of vibration is defined by

$$E = Af^2/497.52 \tag{9.4}$$

where E is energy in acceleration or g units, A is the peak-to-null amplitude, and f is vibration frequency in hertz.

Assuming a constant vibration energy level, increasing vibration frequency reduces amplitude by a significant degree. For example, with an original amplitude equal to 1 grating cycle and a frequency of 10 Hz, doubling the frequency to 20 Hz decreases the amplitude to 0.25 cycle.

There is no evidence to suggest a difference in vibration effects between observer vibration and display vibration, although the former is probably more annoying. There is evidence to suggest that the mechanism used to compensate for vibration changes from eye pursuit at frequencies below 3.4 Hz to fixation of the sine wave nodal points at higher frequencies [22].

The final topic relative to the display environment we will consider is acoustic noise. Here, the data appears totally contradictory. A human engineering guide for image interpretation equipment [13] recommends that the noise level be equal to or less than the value specified by a noise criterion (NC) curve of 45. The NC curves specify the allowable sound pressure level at each of eight sound frequencies (octave bands). The curves allow greater sound pressure values at lower frequencies and decrease the allowable levels as frequency increases. The NC curves are then related to the ability to carry on conversation face-to-face, in a conference, and on the phone. Recommendations are made for various environments. The NC 45 level is equivalent to that found in an engineering or drafting office. Phone conversation is occasionally difficult (because of the noise level). An NC level of 20–30 is specified for an executive office and a level of 30–40 for a medium-size office. The reference [13] also specifies limits for maximum sound levels considered by the Occupational Safety and Health Administration (OSHA) to cause hearing damage.

In a review of 21 studies on the effects of noise on vigilance tasks (which include searching imagery for targets), the authors concluded that nothing is known about the effects of noise on vigilance [23]. The problem is that noise can have different effects depending on such things as content, loudness, and duration. Some studies found that noise degraded vigilance; others found that it had no effect or even enhanced vigilance; results were very specific to the conditions in each study. Generalizations could not be made.

9.2 Observer Characteristics

Virtually all studies of interpretability assess the performance of a sample of observers and report results in terms of the mean of the group. Individuals within the group do differ, however, in a variety of ways. Early research on photointerpreters addressed the issue of selection. What distinguished a good interpreter from a poor interpreter or observer? Could a battery of tests be developed to identify better performing observers? To what degree could performance be improved with vision training? Finally, models of the exploitation process have been developed in terms of the human observer.

9.2.1 Observer Selection

World War II saw considerable research on personnel selection, particularly pilots. Following World War II in the late 1950s, this type of research was extended to photointerpretation. Performance was defined in terms of scores on a photointerpretation test, course grades in a photointerpretation school, and instructor ratings of student proficiency. Predictor variables included tests of visual ability, perception, personality, reasoning, intelligence, and subject matter knowledge related to photointerpretation, and even a test of photo-interpretation. In the typical study, a group of students was given the predictor tests before they entered a class or school and their performance assessed after completion of the course. One of the difficulties of such studies is that a selection process had already occurred prior to the first set of predictor tests. In the military studies, for example, students already had normal vision or better. Further, in an Army study, students were either college graduates or had a General Technical Aptitude score of 100 or greater [24].

In an Air Force study [25], two shadow analysis tests were constructed as a predictor of instructor ratings of recognition and interpretation performance. Figure 9.22 provides an example. Scores on the Army Area Aptitude I test as well as scores on another test of spatial relationships were also used. Results were presented in terms of Z scores (standard deviation units). None of the tests appeared to perform significantly better than any other in identifying the 15 most and the 15 least proficient students (out of 130). A combination of all four tests provided twice the separation of any individual test.

A second large Air Force study [26] used aptitude indices, biographical scores, a variety of perceptual and reasoning tests, and tests of visual perception. A sample of 200 airman trainees was evaluated. A proficiency test involving interpretation of industrial aerial photos was used as the measure of photointerpretation performance. Two aptitude indexes (mechanical and technician specialty) showed the best correlation (0.50 and 0.47) with performance. Correlations for perceptual aptitude tests (e.g., spatial orientation) showed correlations ranging from 0.29 to 0.44. No attempt was apparently made to construct a test battery consisting of multiple tests.

A similar study was performed by the Army using 200 officers and 65 enlisted personnel [24]. Performance on a test requiring interpretation of tactical targets on aerial photography was used as a validity predictor criterion. For officers, a two-test battery consisting of an image orientation test and a test related to military information knowledge was selected. The validity coefficient, however, was only 0.62. For the enlisted sample, the General Technical Aptitude score showed a validity coefficient of 0.60.

The Army and Air Force studies also attempted to predict course grades and incorrect responses. Course grades were predicted with somewhat greater success than was performance; incorrect responses could not be successfully predicted.

In a study using 99 college students [27], a simulated photo image test and a stereo photo identification test were used to predict performance

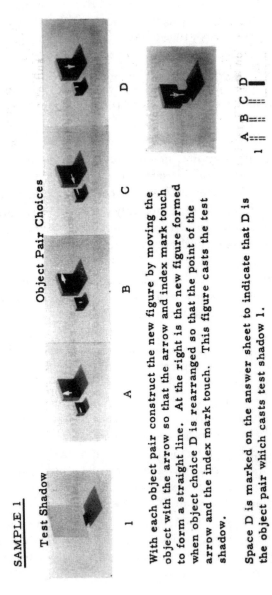

Figure 9.22 Example item from shadow test.

(following training) on a visual search test and a stereo identification test. A validity coefficient (correlation of pre- and post-training test scores) of 0.69 was reported.

In performing search, successive fixation points are based on perception in the peripheral field. A relationship between peripheral vision and search performance might thus be expected. At least three studies demonstrated such a relationship with artificial forms [28–30]. Two other studies attempted to improve peripheral vision through training; results were inconclusive [31,32]. A final study attempted to improve peripheral field size through training as well as to investigate the relationship between field size and search performance on aerial imagery [33]. The same study also presented a 40-hour training program in search techniques to a portion of the imagery analyst sample. Field size was measured using a device that projected a Landolt C in the periphery; the analysts were required to determine the orientation of the gap while maintaining fixation on a central point. Field size improved on the second administration of the test but not universally with further testing or training. Two of seven analysts showed an increase in field size with training; the remainder did not. Field size showed a correlation of 0.72 with search performance (detection completeness). Three of the four analysts receiving the search training program showed significant improvement in search performance. It was concluded that a measure of peripheral acuity might provide a useful addition to other analyst selection tests.

In this same study, two matched versions of the search test were administered, one before peripheral measurement and training and one after the training. The correlation was 0.72 for the seven analysts receiving training and 0.87 for the six who did not receive training. The latter value suggests that search performance is stable over time (up to 15 months in this study).

With multispectral imagery, color vision theoretically becomes important. A significant portion of the population has some form of color deficiency. Deficiencies may exist in the ability to distinguish red and green, blue and yellow, or both. Roughly 7% of the male population and less than 0.5% of the female population are judged to be color deficient [34]. The impact of such deficiencies is not known and may be less than would otherwise be suspected. Dark and light tones as well as spatial details are still distinguishable by people with color discrimination deficiencies. Furthermore, the need to distinguish red from green or blue from yellow is a function of the type of multispectral imagery and the objects of interest. False-color IR imagery tends to appear as blue and red. SWIR composites are often green and yellow. Finally, the ability to manipulate imagery in softcopy may remove any disadvantage of all but the most extreme color deficiency.

9.2.2 Observer Variability

Despite the lack of ability to accurately predict interpreter performance prior to training, it is apparent that analysts differ in ability. As an example, Figure 9.23 shows results of an IR target identification study [35]. The mean of the overall group is 0.68; for the top five performers, it is 0.78, almost a 15% improvement.

In detection studies, the observer variability can be even greater. Figure 9.24 shows results for four observers in one such study [36]. This was a forced choice study in that the observers were presented with a scene and required to indicate whether or not a target was present. The 25% difference in detection rate and the 80% difference in false alarms illustrate the variability in observer detection performance.

Recalling the receiver operating characteristic (ROC) curve from the theory of signal detection (Section 5.2), the difference can be even greater

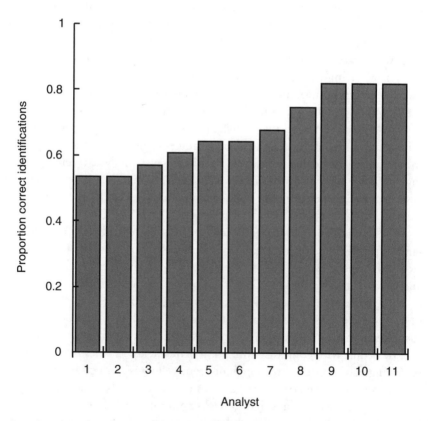

Figure 9.23 Effect of analysts' variability on vehicle identification task. (Data from [35].)

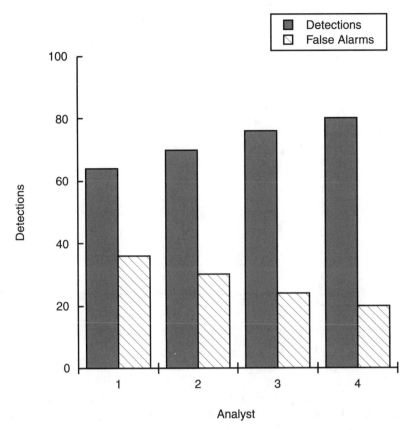

Figure 9.24 Observer variability in forced choice detection study. (Data from [36], © Veridian Systems/Veridian ERIM International Inc.)

in that different observers can be spread out along the ROC curve. Different observers can also exhibit different sensitivity (d'). Where target search is involved, still greater variability can exist. Performance is spread along the FROC curve. It is not uncommon, in the experience of the author, to find one observer whose false alarm rate (number of false alarms) exceeds the mean of the group by as many as six standard deviations.

Finally, as we age, our ability to distinguish contrast differences decreases [13]. Figure 9.25 shows Briggs data for two groups, each with five observers. One group was 39 and younger, the other was 40 and older. The two groups show significant differences in Briggs scores. When observers are aware of their diminished visual capability, however, they can compensate by using magnification and contrast enhancement.

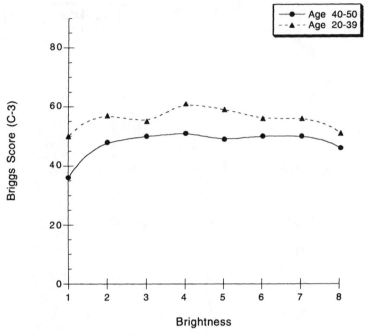

Figure 9.25 Effect of age on Briggs target scores. (Data from [37].)

Subjective responses, of course, also show variability, but the variability is relatively meaningless. With objective data, one can improve performance through selection of observers (or observer results) or training. Subjective data represents performance predictions, and, although they are reasonably accurate in the aggregate, they cannot be used to predict results for individuals.

9.3   Observer Models

In this section, we describe illustrative models dealing with the human visual process. Extraction of information from S&R imagery often begins with a search and detection process. The image is visually scanned in an attempt to locate and identify targets of interest. The search process consists of successive fixations where the eye focuses on a specific point in the image. Movements between fixations are termed "saccades." Fixations are measured in units of time and saccades in angular units. In order for target detection to occur, three things must happen. The target must be fixated, or, if it is sufficiently large, be near a fixation point. The target (and its elements) must be above a detectability threshold. Finally, the target must be recognized or identified as a target.

Search occurs in two modes. In the random or pre-attentive mode, small duration fixations are separated by large saccades. In this mode, nearly unconscious decisions are made regarding whether or not an area or point is of further interest. In the inspection mode, saccades are shorter, reflecting the need to reach a decision on an item of interest. In the context of vision theory, an image is first formed on the receptors of the retina. The image is then decomposed into spatial frequency channels that may or may not be independent. The data in these separate channels is then processed in parallel by the brain and then reassembled to form perception.

The cumulative probability of detection in search can be defined as

$$P_d = 1 - (1 - P_g)^n \qquad (9.5)$$

where P_d the cumulative probability, P_g is the single fixation detection probability, and n is the number of fixations [38].[5] The number of fixations is defined as

$$n = (t - t_r)/t_g \qquad (9.6)$$

where t is the total search time, t_r the time required for the observer to respond to the fixated point (i.e., decide whether or not it is worth further inspection), and t_g is the average fixation time. Although the formulation is simple, it has been shown that rough estimates of the unknowns are not sufficient.

The probability of detecting a target in a single defined fixation P_S is redefined as

$$P_S = 1 - 0.5^{(F_H CR)^3} \qquad (9.7)$$

where F_H is a factor (0–1) designed to account for observer effects such as fatigue or boredom, and CR is defined as the ratio of displayed contrast, C, to threshold contrast, C_T:

$$CR = C/C_T \qquad (9.8)$$

In this model, Blackwell's data [39] is used to define C_T. Separate values are used for photopic and scotopic vision. Presumably they could be averaged

---

5. Equations (9.5) through (9.14) are © SPIE 1991.

in a weighted fashion for the common case in which both levels are required. The form for foveal photopic vision is

$$\log C_{FT} = (0.075 - 1.48 \times 10^{-4}\alpha)(\log L - 2.96 + 1.025\log \alpha)^2 \quad (9.9)$$

where L is the luminance level in fL and α is the target subtense in degrees. The form for parafoveal vision is

$$C_{PFT} = C_{FT}\left[1 + 0.5\left(1 + \frac{\log L}{3}\right)\theta^{(1.5+0.1\log L)}\right] \quad (9.10)$$

where C_{FT} and L are as defined for (9.9), and θ is the angle between the target and the visual axis. Others have argued, however, that detection occurs only with foveal vision. Unless the target is large and uncluttered, direct fixation is required [40]. The value P_S is the value for a single specific fixation. The value will vary for each fixation because contrast will vary across the image. The average single fixation probability P_g is defined by averaging the fixations in the search area. The number of fixations is defined by dividing the size of the image (in degrees) by the average interfixation distance in degrees. Thus,

$$P_g = \frac{1}{mn}\sum_1^m\sum_1^n P_{gs} \quad (9.11)$$

where m and n are the number of fixations in x and y in the field of view and P_S is the single fixation probability. Data from Enoch [19] was used to define fixation times and interfixation angles. Fixation time was defined as

$$t_g = (A_0 + A_1 C_L)\theta_s^{-0.2132} \quad (9.12)$$

where C_L is a measure of the clutter level (0–1) and A_0 and A_1 are constants designed to define a linear relationship between clutter level and fixation time. In the referenced study, A_0 was defined as a value of 0.5782 and A_1 as 1.054. The average interfixation angle was defined as

$$\theta_s = 0.152 t_g^{-3.127} \quad (9.13)$$

Values for t_r and F_H were established empirically using data from another study [41]. This study showed cumulative detection probabilities versus time such that t_r could be defined graphically. There was a dependence on the degree of similarity between targets and clutter. The value of F_H was also dependent on clutter and was defined as

$$F_H = \exp(-5.46C_L^{2.37})$$ (9.14)

The model was shown to correlate well with laboratory data where the background luminance level was greater than 1 fL. The model was subsequently incorporated in a target detection and recognition model called VISDET. It is apparent that results are heavily dependent on the values of t_r and F_H. A subsequent study [42] indicated that the VISDET model overpredicted detection probabilities in a task requiring search for military vehicles in urban scenes. Nonetheless, the model illustrates some of the major factors in modeling the search process. A second search model has been developed in conjunction with the Night Vision and Electronic Sensors Directorate (NVESD) probability of discrimination model discussed in Section 10.1.2. Because it is related, it is discussed in Chapter 10. However, the limitations of the VISDET model also apply to the Night Vision Laboratory (NVL) search model. A key issue in modeling search is the assumption of a random search pattern. In fact, the pattern is often not random. One does not search an urban area for an airfield, for example. Vehicles tend to be found on roads and trails. It is also the case that most search is not systematic and does not cover the total image in a uniform manner. Studies of search in real-world conditions tend to show nonsystematic patterns. For these reasons, accurate models of the search process may not be possible in real-world conditions.

Modeling of detection thresholds is a more tractable problem. An example is provided by the Barten model [43] used in the NEMA/DICOM display calibration function described in the previous chapter. In the VISDET model, the threshold contrast required for detection was based on Blackwell's data. The threshold value included only the luminance of the target (a small disc) and the luminance of the background. The Barten model includes the effects of image noise, internal noise (photon and neural), filtering of low spatial frequencies in the ganglion cells (lateral inhibition), the optical MTF of the eye, and target size. The model is currently defined for binocular viewing as [44]

$$1/m_t(u) = \frac{M_{opt}(u)/k}{\sqrt{\frac{2}{T}\left(\frac{1}{x_0^2} + \frac{1}{x_{max}^2} + \frac{u^2}{N_{max}^2}\right)\left(\frac{1}{\eta pE} + \frac{\phi_o}{1 - e^{-(u/u_o)^2}}\right)}} \qquad (9.15)$$

where $m_t(u)$ is the threshold modulation at frequency u, $M_{opt}(u)$ is the optical MTF of the eye at frequency u, k is a constant (3.0), and T is the integration time of the eye (0.1 sec). For monocular viewing, the value 2 under the square root sign is replaced by a 4.

The eye has a limited ability to integrate in the spatial domain. The expression

$$X = \left\{1/x_o^2 + 1/x_{max}^2 + (u/N_{max})^2\right\}^{-0.5} \qquad (9.16)$$

is used where X_o is the angular size of the object, X_{max} is the maximum angular size of the integration area, and N is the maximum number of cycles over which integration can occur. A value of 12 degrees is used for X_{max} and 15 cycles for N_{max}.

The term $1/\eta pE$ defines photon noise where η is total quantum efficiency (0.03) and p is a photon conversion factor (which varies as a function of the light source) to convert light units to units of photon flux density entering the eye(s). For a P4 white phosphor, a value of 1.24 per Troland[6] (Td) is provided for photopic viewing. This constant p times the illuminance E (measured in Td) and the quantum efficiency η defines the photon flux J passing through the cornea. E is the illuminance of the eye defined as

$$E = \frac{\pi d^2}{4} L \qquad (9.17)$$

where d is the diameter of the eye pupil (in mm) and L is the luminance of the target (in cd/m^2). E is measured in trolands. Neural noise is defined by ϕ_o and has an estimated value of 3×10^{-8} sec deg^2.

---

6. Troland is a measure of retinal illuminance defined as scene luminance (cd/m^2) times the effective pupil area of the eye in millimeters squared. The effective pupil area is the actual area times an effectivity ratio based on the pupil diameter. The effectivity ratio varies from 0.989 to 0.668 over a pupil diameter range of 1 to 6 mm and provides a correction for the decrease in luminance sensitivity at increasing distance from the center of the pupil.

The lateral inhibition process attenuates low spatial frequency components. The MTF is defined as

$$M_{lat}(u) = \sqrt{1 - e^{-(u/u_o)^2}} \qquad (9.18)$$

where u is the angular spatial frequency and u_o is the frequency, defined as 7 cycles/degree, at which lateral inhibition no longer has an effect. Finally, the optical MTF of the eye is defined as

$$MTF_{opt}(u) = e^{-\pi^2 \sigma^2 u^2} \qquad (9.19)$$

The term σ is the standard deviation of the line spread function due to the convolution of various effects in the eye that contribute to the total effect. It is defined as

$$\sigma = \sqrt{\sigma_o^2 + (C_{ab}d)^2} \qquad (9.20)$$

where σ_o is a constant (0.5 arc min), C_{ab} is another constant (0.08 arc min/mm) and d is the pupil diameter.

The model was shown to be a good fit to data from 14 different studies. The studies covered a range of luminance values, spatial frequencies, field sizes, light types (monochromatic and white), and with and without noise. At this point in the process, the target has been fixated and detected, but must still be recognized as a target or item of interest. We will use the term "recognition" here to refer to the process by which a name is applied to a stimulus. A stimulus may be recognized as a military object, a vehicle, a tank, or an M-60 tank. A variety of recognition models have been postulated. Broadly, they require the detection of a set of features such as edges, shapes, and textures, and comparison of the feature set to some type of stored (in memory) "feature dictionary." In searching for a particular target, it can be argued that recognition is a change detection process, entailing a cross-correlation between the expected image and the actual image at each fixation point [45]. This comparison may be serial or parallel [46]. Parallel search is considered to be almost automatic and to require little attention. Searching for highway warning signs might be considered an example. When we look at a stop sign, we do not consciously compare it to all possible sign shapes and colors. On the other hand, searching S&R imagery for vehicles in a wooded environment requires serial processing each time a possible edge is detected.

Figure 9.26 provides a conceptual description of the serial recognition process [47] developed by Biderman. Biderman develops a theory of recognition by components; those components being simple convex shapes such as cylinders, blocks, spheres, and wedges. Edges define those shapes. Nonaccidental properties are those that provide an indication of the object in three-dimensional space based on a two-dimensional representation. For example, consider the cube and box shown in Figure 9.27. Both are perceived as having flat sides and parallel surfaces. The cube is perceived as having the same dimensions in each plane. Parsing (separation into parts) occurs at points or regions where the surface is concave, typically where one simple shape is joined to another. Figure 9.28 shows two examples. Both shapes are perceived as two parts or components. Once the components are defined, they can be compared to known object representations. This then leads to object recognition.

The simple components were called geons and it was postulated that some 36 geons could define all "count nouns," although the number of 36

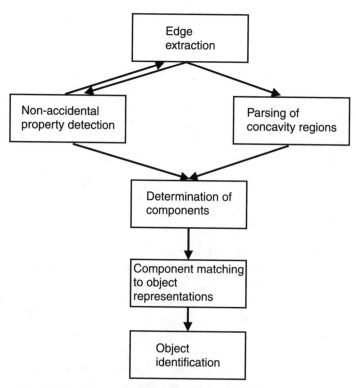

Figure 9.26 Conceptual recognition model.

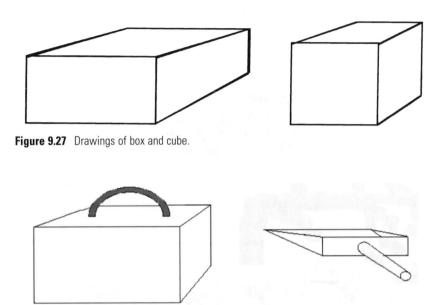

Figure 9.27 Drawings of box and cube.

Figure 9.28 Object components.

is not critical to the theory. "Count nouns" were defined as objects with specified boundaries to which an indefinite article and number can be applied: one tank or three tanks, one airplane or 10 airplanes. Noncount or mass nouns do not take an article or number (e.g., snow, water, sand). The theory also states that recognition occurs at a higher level than suggested by the typical recognition hierarchy. For example, tank as opposed to vehicle, house as opposed to building. The theory also states that surface characteristics such as color or tone and texture play a secondary role. Biderman provided an estimate of some 30,000 easily discriminable objects (3000 categories such as chair and elephant with 10 discernable examples each). The number of possible two-geon objects was estimated at 75,000. With the possibility of three geon objects, the number rises to 1.54 million. Thus, the 36 geons are more than sufficient to account for the number of discriminable objects. It was also shown that some objects (an airplane for example) can have many more than three components, although the number of unique geons may be less. The tank shown in Figure 9.29 can be broken into perhaps three major geons. They can be described as a cylinder (gun), a box (hull), and some type of elliptical cylinder (turret). The separation of the hull and the tracks can also be argued. Additional geons such as the hatches (cylinders) add detail that may be used to identify the tank model.

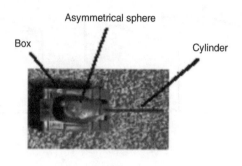

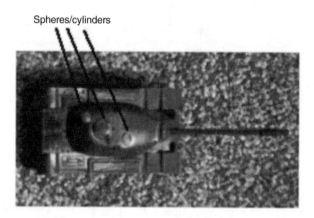

Figure 9.29 Tank image components.

In Biderman's paper, data from several reaction time studies were presented to support the geon recognition hypothesis. Data was also presented to show the effect of deleting portions of the target outline. If the concave vertices are eliminated, recognition cannot occur. The geon theory is interesting in several ways. It is similar to speech perception that indicates that only 44 phonemes (primitive elements) are required to code all of the words in the English language. It is not inconsistent with the target recognition data presented in Chapter 7. Finally, it suggests that recognition of objects on SAR imagery may be more difficult because of the lack of obvious or continuous edges as shown in Figure 9.30.

The geon theory of recognition was proposed for objects composed of shape primitives. Natural features are more often described or identified in terms of textural features, at least on smaller scale imagery where individual elements are not evident. A forest canopy is seen as a pattern of rounded blobs

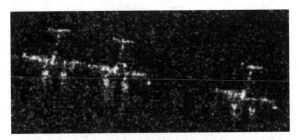

Figure 9.30 SAR image showing lack of defined edges (courtesy of TESAR Program).

and grass as a more uniform pattern. At larger scale, individual trees may be seen and decomposed into geons (cone and cylinder for most coniferous trees). A term called "Texton" has been proposed as the basic unit of pre-attentive texture perception [48]. Textons are defined in terms of heuristics. Textural features are aggregates or patterns of small features. It is the pattern or texture that characterizes the feature, rather than the individual elements. Textons are defined as elongated blobs with specific colors, orientation, shape, and size. Textons can also be defined in terms of end-of-line segments and line crossings, although not normally for aerial imagery. Finally, the positional relationship between textons is ignored in pre-attentive vision. SAR clutter (Figure 9.30) forms an image example. The texton theory is consistent with the random or pre-attentive mode of search.

9.4 Summary

In this chapter, the effects of the display on information extraction were shown. Softcopy outperforms hardcopy only when the display meets certain performance criteria and the display environment does not degrade performance. High levels of ambient light, in particular, can result in interpretability losses of 0.5 NIIRS or greater. Attempts to predict success in image interpretation have been met with limited success. Tests of general and specific technical aptitude have been somewhat more successful than tests of perceptual ability. Vision tests have not been successful, probably because students entering image interpretation training have already gone through a vision screening process. The exception is peripheral acuity; there is evidence that persons with larger visual fields perform better in search tests than those with smaller fields. Attempts to enlarge visual fields through training, as well as providing training in the search process, have been met with limited success. Individual observers differ in performance, perhaps partly because of ability and partly because of their decision threshold. Older observers show less contrast sensitivity than younger observers, but they can compensate with magnification and contrast enhancement.

Many S&R models treat the observer as a black box and do not model the observer. A variety of models, however, have been developed to explain the process of search and recognition. These models generally fall in three categories: those dealing with search, those dealing with threshold contrast and spatial detection, and those dealing with object recognition. These models are useful in providing an understanding of the S&R information extraction process.

References

[1] Leachtenauer, J., et al., "The General Image Quality Equation," *Applied Optics,* November 1997, pp. 8322–8328.

[2] Latshaw, G. L., P. L. Zuzelo, and S. J. Briggs, "Tactical Photointerpreter Evaluation of Hardcopy and Softcopy Imagery," in *Airborne Reconnaissance III: Collection and Exploitation of Reconnaissance Data,* SPIE, Vol. 137, 1978, pp. 179–182.

[3] Leachtenauer, J., K. Daniel, and T. Vogl, "Digitizing Satellite Imagery," *Photogrammetric Engineering and Remote Sensing,* Vol. LXIV, No. 1, January 1998, pp. 29–34.

[4] Leachtenauer, J. C., and N. L. Salvaggio, "NIIRS Prediction: Use of the Briggs Target," *ASPRS/ASCM Annual Convention and Exhibition Technical papers,* Volume 1, *Remote Sensing and Photogrammetry,* American Society for Photogrammetry and remote Sensing and American Congress on Surveying and Mapping, Baltimore, Maryland, April 22–25, 1996, pp. 282–291.

[5] Leachtenauer, J., A. Biache, and G. Garney, "Effect of Ambient Lighting and Monitor Calibration on Softcopy Image Interpretability," *Final Program and Proceedings, IS&T PICS Conference,* IS&T, Savannah, Georgia, 1999, pp. 179–183.

[6] National Information Display Laboratory, *Test Procedures for Evaluation of CRT Display Monitors,* Princeton, NJ: David Sarnoff Research Center, 1991.

[7] Video Electronics Standards Association, *Flat Panel Display Measurements Standard (Proposal),* San Jose, CA: VESA, 1997.

[8] Briggs, S. J., *Manual Digital Test Target BTP#4,* D180-25066-1, Seattle, WA: Boeing Aerospace Company, 1979.

[9] Briggs, S. J., *Digital Display Test Target Development,* D180-15860-1, Seattle, WA: Boeing Aerospace Company, 1977.

[10] Leachtenauer, J., and N. Salvaggio, "Color Monitor Calibration for Display of Multispectral Imagery," *Society for Information Display, International Symposium, Digest of Technical Papers,* Boston, Massachusetts, May 13–15, 1997, pp. 1037–1040.

[11] Wulfeck, J., A. Weisz, and M. Raben, *Vision in Military Aviation,* WADC-TR-58-399, Wright-Patterson Air Force Base, OH: Wright Air Development Center, 1958.

[12]   NEMA/DICOM, *Grayscale Standard Display Function*, PS 3.14-1998, Arlington, VA: National Electrical Manufacturers Association, January 1998.

[13]   Farrell, R. J., and J. M. Booth, *Design Handbook for Imagery Interpretation Equipment*, Seattle, WA: Boeing Aerospace Company, 1984.

[14]   National Information Display Laboratory, *A Survey of Twelve Color CRT Monitors*, Princeton, NJ: David Sarnoff Research Center, 1997.

[15]   Leachtenauer, J., G. Garney, and A. Biache, "Contrast Modulation—How Much Is Enough?" *Final Program and Proceedings, IS&T PICS Conference*, IS&T, Portland, Oregon, 2000, pp. 130–134.

[16]   Leachtenauer, J., A. Biache, and G. Garney, "Effects of Pixel Density on Softcopy Image Interpretability," *Final Program and Proceedings, IS&T PICS Conference*, IS&T, Savannah, Georgia, 1999, pp. 184–188.

[17]   Keller, P. A., *Electronic Display Measurement: Concepts, Techniques, and Instrumentation*, New York, NY: Wiley, 1997.

[18]   Ludveigh, E., and J. W. Miller, "Study of Visual Acuity During the Ocular Pursuit of Moving Test Objects—I. Introduction," *JOSA*, Vol. 65, 1958, pp. 799–802.

[19]   Enoch, J. M., "Effect of the Size of a Complex Display Upon Visual Search," *JOSA*, Vol. 49, 1959, pp. 280–286.

[20]   Erickson, R. A., et al., *Resolution of Moving Imagery on Television: Experiment and Application*, China Lake, CA: Naval Weapons Center, 1974.

[21]   Rhodes, F., and H. C. Self, *The Effect of Direction and Speed of Image Motion Upon Target Detection With Side-Looking Radar*, AMRL-TDR-64-45, Wright-Patterson Air Force Base, OH: Aerospace Medical Research Laboratory, 1964.

[22]   Huddleston, J. H. F., "Tracking Performance on a Visual Display Apparently Vibrating at One to Ten Hertz," *Journal of Applied Psychology*, Vol. 54, No. 5, 1970, pp. 401–408.

[23]   Kolega, H. S., and J. Brinkman, "Noise and Vigilance: An Evaluative Review," *Human Factors*, Vol. 28, No. 4, 1986, pp. 465–482.

[24]   Martinek, H., R. Sadacca, and L. Burke, *Development of a Selection Battery for Army Image Interpreters*, TRN 1143, Washington, DC: U.S. Army Personnel Research Office, 1965.

[25]   Reyna, L. J., P. Nogee, and S. R Mayer, *A Study of Two Tests for Discrimination of Proficient Photo-Interpreter Students*, Technical Note 114, Boston, MA: Boston University Optical Research Laboratory, 1954.

[26]   Meyer, J. K., and R. E. Miller, *Validity of Photo Interpreter Predictors for Test and Training Criteria*, WADD-TN-60-45, Lackland Air Force Base, TX: Air Research and Development Command, 1960.

[27]   Avery, T. E., and H. Burkhart, "Screening Tests for Rating Photointerpreters," *Photogrammetric Engineering*, Vol. 34, May 1968, pp. 476–482.

[28]   Erickson, R. A., "Relation Between Visual Search Time and Peripheral Visual Acuity," *Human Factors*, Vol. 6, No. 2, 1964, pp. 165–177.

[29]   Johnston, D. M., "Search Performance as a Function of Peripheral Acuity," *Human Factors*, Vol. 7, No. 5, 1965, pp. 527–535.

[30]   Boynton, R. M., C. L. Elworth, and R. M. Palmer, *Laboratory Studies Pertaining to Visual Air Reconnaissance,* Part III, WADC-TR-55-304, Wright-Patterson Air Force Base, OH: Wright Air Development Center, 1958.

[31]   Low, F. *Effect of Training on Acuity of Peripheral Vision*, PB 50339, Washington, DC: Civil Aeronautics Administration, 1946.

[32]   Crannel, C. W., and J. M. Christensen, *Expansion of the Visual Form Field by Perimeter Training*, WADC-TR-55-368, Wright-Patterson Patterson Air Force Base, OH: Wright Air Development Center, 1958.

[33]   Leachtenauer, J. C., "Peripheral Acuity and Photointerpretation Performance," *Human Factors*, Vol. 20, No. 5., 1978, pp. 537–551.

[34]   Judd, D. B., and G. Wyszecki, *Color in Business, Science and Industry*, New York, NY: Wiley, 1963.

[35]   Driggers, R., et al., "Sensor Performance Conversions for Infrared Target Acquisition and Intelligence-Surveillance-Reconnaissance [ISR] Imaging Sensors," *Applied Optics*, Vol. 38, No. 31, 1999, pp. 5936–5943.

[36]   Voas, R. B., W. Frizzell, and W. Zink,, *Evaluation of Methods for Studies Using the Psychophysical Theory of Signal Detection*, Washington, DC: Environmental Research Institute of Michigan, 1987.

[37]   Leachtenauer, J. C., G. Garney, and A. Biache, "Effects of Monitor Calibration on Imagery Interpretability," *Final Program and Proceedings, IS&T PICS Conference*, IS&T, Portland, Oregon, 2000, pp. 124–129.

[38]   Waldman, G., J. Wooton, and G. Hobson, "Visual Detection With Search: An Empirical Model," *IEEE Transactions on Systems, Man, and Cybernetics*, Vol. 21, No. 3, 1991, pp. 596–606.

[39]   Blackwell, H. R., "Contrast Thresholds of the Human Eye," *JOSA*, Vol. 36, 1946, pp. 624–643.

[40]   Biederman, I., et al., "Detecting the Unexpected in Photointerpretation," *Human Factors*, Vol. 23, No. 2, 1981, pp. 153–164.

[41]   Bloomfield, J. R., "Visual Search in Complex Fields: Size Differences Between Target Disc and Surrounding Discs," *Human Factors*, Vol. 14, No. 2, 1972, pp. 139–148.

[42]   Toet, A., P. Bijl, and M. Valeton, "Test of Three Visual Search and Detection Models," in W. R. Watkins, D. Clement, and W. R. Reynolds (eds.), *Proc. SPIE Vol. 3699, Targets and Backgrounds: Characterization and Representation*, 1999, pp. 323–334.

[43]   Barten, P.G., "Physical Model for the Contrast Sensitivity of the Human Eye," in *Human Vision, Visual Processing, and Digital Display III*, SPIE Vol. 1666, 1992, pp. 57–72.

[44]   Barten, P.G., *Contrast Sensitivity of the Human Eye and Its Effects on Image Quality*, Bellingham, WA: SPIE Optical Engineering Press, 1999.

[45] Caelli, T., and G. Moraglia, "On the Detection of Signals Embedded in Natural Scenes," *Perception and Psychophysics*, Vol. 39, No. 2, 1986, pp. 87–95.

[46] Schneider, W., and R. M. Shiffren, "Controlled and Automatic Human Information Processing: I. Detection," *Psychological Review*, Vol. 84, No. 1, 1977, pp. 1–66.

[47] Biederman, I., "Recognition by Components: A Theory of Human Image Understanding," *Psychological Review*, Vol. 94, No. 2, 1987, pp. 115–147.

[48] Julesz, B., and R. J. Bergen, "Textons, the Fundamental Elements in Pre-attentive Vision and Perception of Textures," *The Bell System Technical Journal*, Vol. 62, No. 6, 1983, pp. 1619–1645.

10

Performance Prediction Models

In previous chapters, we discussed both information extraction measures and information extraction performance predictors. We also showed the impact of elements of the image chain on the performance of S&R systems where performance was defined in terms of information extraction. In this chapter, we review candidate performance prediction models. We define a model as a set of procedures and processes that predict the performance of a particular system or system design operating under a specified set of conditions. Because there are a variety of ways in which performance prediction models can be classified, we have divided them into parameter-based models and image-based models. Image-based models require the availability of imagery. Parameter-based models can be exercised given knowledge of the system design and operating parameters; imagery is not required. Most parameter-based models, however, can also be run using image measurements.

Broadly, performance prediction models have been developed for four classes of imaging systems: SAR, downward-looking visible and IR, battlefield target acquisition (low elevation angle visible and IR), and video. The primary performance metrics are NIIRS, probability of task performance, and subjective quality ratings. Table 10.1 categorizes the models discussed in this chapter. The models are described in sufficient detail to provide reader understanding; model validation data is provided when available. Finally, potential extensions to the models are discussed, as is the need for further development.

Table 10.1
Performance Prediction Models

Model	Type	Application	Systems	Metric
General image quality equation	Parameter	Downward-looking	Visible/IR	NIIRS
NVESD	Parameter	Target acquisition	Visible/IR	Probability of discrimination
Physique	Parameter	Downward-looking	Any*	NIIRS
Radar threshold Quality factor	Parameter	Downward-looking	SAR	Probability of recognition
Other SAR	Parameter	Downward-looking	SAR	Image quality
Image quality metric	Image	Downward-looking	Any	NIIRS
Sarnoff JND	Image	Any	Any**	NIIRS, subjective quality
Other visible	Image	Any	Video	Subjective quality

*The Physique models developed for visible, IR, and SAR systems have no published validation data.
**The Sarnoff JND model has been demonstrated only with visible imagery and video.

10.1 Parameter-Based Models

Parameter-based models are of primary use to the system designer. If the necessary input parameter data can be defined, the model can be used to predict the performance of a system that has not yet been built. Similarly, given a performance goal, system design trades can be performed. The GIQE, the first of the parameter-based models to be reviewed, has been used as a tool in UAV sensor design [1] and extensively validated, but over a somewhat limited set of conditions. Similarly, the Night Vision and Electronic Sensors Directorate (NVESD) probability of discrimination model [2] has been used in the development of battlefield target acquisition systems.

Physique is a model that predicts information, which, in turn, can be related to a performance metric. Models have been developed for visible and SAR systems [3,4], but no validation data has been published. Only the visible model will be discussed.

Finally, the radar threshold quality factor (RTQF) model [5] is an SAR performance prediction model based on Charmin and Olin's threshold quality factor (TQF) [6]. The RTQF underwent a small validation effort. Although modeling of SAR systems is well advanced in the context of modeling the appearance of images, there is virtually no data relating quality to performance of human observers. The modeling focus appears to have been on automatic target recognition.

10.1.1 General Image Quality Equation

The GIQE was developed in the 1980s, but was not formally released to the UAV development community until December 1994 [1]. Both the Predator and Global Hawk EO sensors were specified in terms of NIIRS performance [7].[1] Consequently, it was necessary for developers to have available some tool to predict NIIRS as a function of their system design.

The original release of the GIQE (Version 3.0) predicted visible NIIRS as a function of GSD (scale and resolution), sharpness (edge sharpness), and signal-to-noise ratio. The ability to predict IR NIIRS was not available in the original release, although suggestions for modification to allow at least relative comparison of alternative designs were provided. The validation of Version 3.0 was performed on hardcopy visible EO imagery.

Subsequent to the release of GIQE Version 3.0, an extensive update and re-validation effort was undertaken [8,9]. A sample of 359 visible and 372 IR images was available for this effort. As a result of this effort, separate equations were developed for the visible and IR (the visible and IR NIIRS are different) and slight revisions were made to the original form of the GIQE.

The conceptual form of the GIQE is shown in Figure 10.1. Version 4.0 of the GIQE is defined for visible imagery as

$$NIIRS = 10.251 + a \log_{10} GSD_{GM} + b \log_{10} RER_{GM}$$
$$+ 0.656 H_{GM} - 0.344 (G/SNR) \qquad (10.1)$$

and for IR imagery as

$$NIIRS = 10.751 + a \log_{10} GSD_{GM} + b \log_{10} RER_{GM}$$
$$+ 0.656 H_{GM} - 0.344 (G/SNR) \qquad (10.2)$$

where GSD_{GM} = geometric mean of ground-sampled distance (in inches), RER_{GM} = geometric mean of the normalized relative edge response, H_{GM} = geometric mean height of overshoot due to MTFC, G = noise gain due to MTFC, SNR = signal-to-noise ratio, a = 3.32 if $RER \geq 0.9$ and 3.16 if $RER < 0.9$, and b = 1.559 if $RER \geq 0.9$ and 2.817 if $RER < 0.9$.

---

1. The existence of an obsolete NATO IIRS resulted in some initial confusion in relating requirements to design parameters. The issue has since been resolved.

GIQE

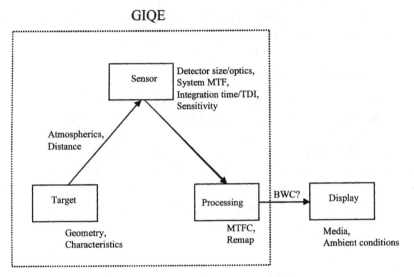

Figure 10.1 GIQE conceptual model.

Target effects (aspect, size, contrast) are captured in the GSD term and potentially the SNR term. For most applications, however, a standard target contrast is assumed. Atmospheric effects are captured in the SNR term, although path length also affects GSD. Sensor effects are handled with the GSD and MTF-related (RER and to a lesser degree, H and G) terms. Processing effects include MTFC and remap (DRA and TTC). Remap is assumed to be optimized, and a hardcopy display is assumed. The effects of bandwidth compression are not included.

10.1.1.1 RER

Turning to the definition of the individual components, RER is the slope of the normalized edge response. It is measured between two points one-half pixel from an edge, as shown in Figure 10.2. The edge response is normalized to the range of 0 to 1. The edge response is measured in the x and y axis and the geometric mean computed.

The normalized edge response is calculated as

$$ERx(d) = 0.5 + \frac{1}{\pi} \int_{o}^{(nOptcutx)} \left[\frac{SystemX(\xi)}{\xi} \times \sin(2\pi\xi d) \right] d\xi \quad \text{X-axis} \quad (10.3)$$

$$ERy(d) = 0.5 + \frac{1}{\pi} \int_{o}^{(nOptcuty)} \left[\frac{SystemY(\xi)}{\xi} \times \sin(2\pi\xi d) \right] d\xi \quad \text{Y-axis} \quad (10.4)$$

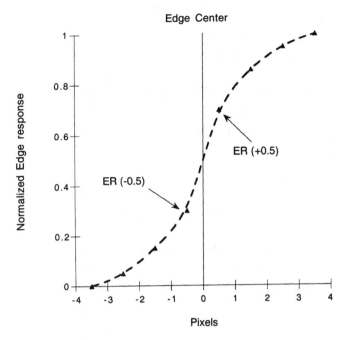

Figure 10.2 RER measurement.

where *nOptcutx, nOptcuty* are the normalized (to the effective sample spacing) optics cutoffs in the x and y directions, *SystemX* and *SystemY* are the system MTFs in the X and Y direction, *d* is the response position from the center of a horizontal pixel, and ξ is the spatial frequency in cycles per sample spacing.[2] The system MTF is normalized to the effective pixel spacing after decimation or aggregation.

Given the values of ER, RER is the response difference between the points +0.5 and −0.5 pixels from the edge:

$$RER_X = ER_X(0.5) - ER_x(-0.5)$$
$$RER_Y = ER_Y(0.5) - ER_Y(-0.5) \qquad (10.5)$$

The geometric mean is defined as

$$RER_{GM} = \sqrt{RER_X \times RER_Y} \qquad (10.6)$$

---

2. The symbol ξ is used in this book to denote spatial frequency in the GIQE. The original reference used v.

The system MTF is required for calculation (as opposed to measurement) of RER. The system MTF is the cascaded value of all the component MTFs. These typically include:

- Optics (design, fabrication, alignment, defocus);
- Environmental wavefront error;
- Boundary layer wavefront error;
- Atmospheric dispersion and turbulence;
- Detector;
- Aperture, clocking, carrier diffusion, charge transfer efficiency;
- Modulation transfer function compensation.

Other components can be added as required. The system MTF is defined over the spatial frequency range from 0 to the maximum passed by the normalized optics aperture defined as

$$p/(\lambda\, f N) \tag{10.7}$$

where p is the effective sample spacing in millimeters after aggregation or decimation, λ is the minimum wavelength (mm), and $f N$ (f number) is the relative aperture of the optics.

10.1.1.2 GSD

Ground-sampled distance is simply the projection of the detector pitch to the ground. Version 3.0 computed GSD in the plane orthogonal to the line of sight. Version 4.0 computes GSD in the ground plane. The change was made on the basis of empirical results. Use of the ground plane projection resulted in a significantly better fit to the data. This topic will be addressed in more detail in the subsequent discussion of GIQE limitations.

GSD is measured in the x and y axes of the detector array perpendicular to the line of sight. If necessary, the pixel pitch dimension is corrected for aggregation or decimation. GSD is defined as

$$\text{GSD} = \{(\text{pixel pitch/focal length}) \times \text{slant range}\}\, \cos(\text{look angle}) \tag{10.8}$$

where linear dimensions are in inches and look angle is in degrees. Look angle is the angle between the ground plane and the line of sight from the ground to the sensor.

The geometric mean is computed for use in the model:

$$GSD_{GM} = \sqrt{(GSD_X \times GSD_Y)} \tag{10.9}$$

If the cross-scan and along-scan directions are not orthogonal on the ground, the sin of the angle (α) between the two directions must be used

$$GSD_{GM} = \sqrt{(GSD_X \times GSD_Y \times \sin\alpha)} \tag{10.10}$$

10.1.1.3 Edge Overshoot

Edge overshoot (H) is the height of the overshoot across an edge resulting from MTFC. It is the value of the peak normalized edge response in the region 1.0 to 3.0 pixels from the edge unless the edge is monotonically increasing over this range. In that case, it is defined at 1.25 pixels from the edge. Figure 10.3 shows the computation. The geometric mean of H calculated in the x and y dimensions is used.

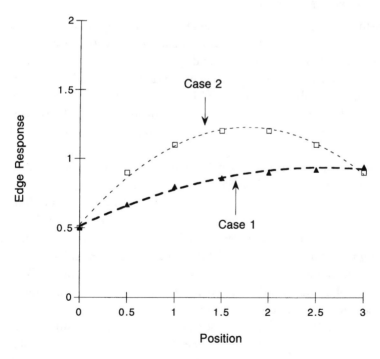

Figure 10.3 Computation of H (edge overshoot).

10.1.1.4 Noise Gain

Noise gain is the increase in noise resulting from MTFC. It is calculated by

$$G = \sqrt{\left(\Sigma_{i=1}^{M} \Sigma_{i=1}^{N} (Kernel_{ij})^2 \right)} \qquad (10.11)$$

where the sum of the squares of the kernel elements is computed and the square root of the sum taken. If MTFC is not applied, a value of 1 for G is used. For the kernel:

$$MTFC = \begin{matrix} -.0732 & -.3536 & -.0732 \\ -.3536 & 2.707 & -.3536 \\ -.0732 & -.3536 & -.0732 \end{matrix}$$

the value of G is 2.8.

10.1.1.5 Signal-to-Noise Ratio

Signal-to-noise-ratio is the ratio of the DC differential scene signal level (electrons) to RMS electrons of noise. It is calculated across the spectral band of interest after calibration, sampling, and aggregation, and before MTFC. DC differential scene radiance is the difference in detector output signals between two extended Lambertian surfaces with different reflectances. A known target and background reflectance can be used, or some nominal assumed surfaces can be used. Commonly, flat surfaces with reflectances of 7% and 15% are used.

Signal calculation requires definition of scene irradiance and path radiance and transmittance. This data can be defined by an atmospheric model such as MODTRAN [10]. MODTRAN was used for GIQE Version 4.0. At the sensor, calculation of focal plane irradiance requires knowledge of the relative aperture and the transmission of the optics. Definition of the focal plane signal requires specification of the detector quantum efficiency, detector size, integration time, and number of time delay integration (TDI) stages. Noise contributions include signal-dependent or photon noise and dark current readout, quantization, and nonuniformity noise.

For IR systems, additional radiance due to internal sources must be considered. Differential scene radiance needs to be calculated in terms of scene emissivity, reflectance, and temperature. Typical values for laboratory blackbody sources are an emissivity of 1, a reflectance of 0, an average temperature of 280 degrees K, and a temperature difference of 2 degrees K.

The computation for SNR used in Version 4.0 of the GIQE was developed by Eisman and Engle [11] and is defined as

$$SNR = \frac{N_{HI} - N_{LOW}}{\sqrt{N_{HI} + \dfrac{J_d N_x N_y w_x w_y T_d}{q} + \sigma_2^n + \dfrac{1}{12}\left[\dfrac{N_{max} - N_{min}}{2^b - 1}\right]}} \qquad (10.12)$$

where

$$N_{HI} = \max(N_{tgt}, N_{bkg}) \qquad (10.13)$$

$$N_{LOW} = \min(N_{tgt}, N_{bkg}) \qquad (10.14)$$

$$N_{tgt} = \frac{\pi T_d N_{TDI} N_x N_y w_x w_y}{4hc(fN)^2}\left\{L_{stray} + \int_{\lambda_{min}}^{\lambda_{max}} \lambda\eta(\lambda)[\varepsilon_{opt}L_{BB}(\lambda, T_{opt})\right.$$
$$\left. + \tau_{opt}(\lambda)\tau_{atm}(\lambda)L_{tgt}(\lambda) + \tau_{opt}(\lambda)L_{path}(\lambda)]d\lambda\right\} \qquad (10.15)$$

$$N_{bkg} = \frac{\pi T_d N_{TDI} N_x N_y w_x w_y}{4hc(fN)^2}\left\{L_{stray} + \int_{\lambda_{min}}^{\lambda_{max}} \lambda\eta(\lambda)[\varepsilon_{opt}L_{BB}(\lambda, T_{opt})\right.$$
$$\left. + \tau_{opt}(\lambda)\tau_{atm}(\lambda)L_{bkg}(\lambda) + \tau_{opt}(\lambda)L_{path}(\lambda)]d\lambda\right\} \qquad (10.16)$$

$$L_{tgt}(\lambda) = \left[1 - \rho_{tgt}(\lambda)\right]L_{BB}(\lambda, T_{tgt}) + \rho_{tgt}(\lambda)\left[\frac{E_{direct}(\lambda)\cos z}{\pi} + L_{diffuse}(\lambda)\right] \quad (10.17)$$

$$L_{bkg}(\lambda) = \left[1 - \rho_{bkg}(\lambda)\right]L_{BB}(\lambda, T_{bkg}) + \rho_{bkg}(\lambda)\left[\frac{E_{direct}(\lambda)\cos z}{\pi} + L_{diffuse}(\lambda)\right] (10.18)$$

where

J_d = detector dark current density (μA(cm^2))

N_{TDI} = time-delay-integration stages

N_x, N_y = aggregation in x and y

$w_x w_y$ = detector width in x and y (μm)

T_d = dwell time (msec)

q = charge of an electron (1.6×10^{-19} coulombs)

σ_n = detector RMS noise (electrons)

N_{max} = quantizer maximum

N_{min} = quantizer minimum

b = quantizer digitizing levels

N_{tgt} = target signal level (electrons)

N_{bkg} = background signal level (electrons)

h = Planck's constant = $(6.6256 \pm 0.0005) \times 10^{-34}$ W sec^2

c = velocity of light = $(2.997925 \pm 0.000003) \times 10^{10}$ cm sec^{-1}

fN = optics f-number

L_{stray} = in-band stray radiance (W/m^{3}sr)

λ = wavelength (μ)

$\eta(\lambda)$ = detector quantum efficiency

ε_{opt} = effective optics emissivity

L_{BB} = blackbody function as defined in (10.19)

T_{opt} = effective optics temperature (K)

$\tau_{opt}(\lambda)$ = optical train transmission

$\tau_{atm}(\lambda)$ = atmospheric path transmittance

$L_{tgt}(\lambda)$ = target radiance (W/m^{3}sr)

$L_{bkg}(\lambda)$ = background radiance (W/m^{3}sr)

$L_{path}(\lambda)$ = atmospheric path radiance (W/m^{3}sr)

$\rho_{tgt}(\lambda)$ = target reflectance

$\rho_{bkg}(\lambda)$ = background reflectance

$E_{direct}(\lambda)$ = direct source irradiance (W/m^3)

z = source zenith angle (degrees)

$L_{diffuse}(\lambda)$ = diffuse downwelling radiance (W/m^{3}sr)

T_{tgt}, T_{bkg} = target, background temperature (K)

$N_{tgt} N_{bkg}$ = target, background signal level (electrons)

$$L_{BB}(\lambda, T) = \frac{c_1}{\lambda^5} \frac{1}{e^{c_2/\lambda T} - 1} \tag{10.19}$$

where

λ = wavelength (μ)

T = absolute temperature (K)

h = Planck's constant = $(6.6256 \pm 0.0005) \times 10^{-34}$Wsec^2

$c_1 = 2\pi hc^2$ = first radiation constant = $(3.7415 \pm 0.0003) \times 10^4$Wsec^2

c = velocity of light = $(2.997925 \pm 0.00003) \times 10^{10}$cm sec^{-1}

$c_2 = ch/k$ = second radiation constant = $(1.43879 \pm 0.00019) \times 10\mu$K

k = Boltzman's constant = $(1.38054 \pm 0.00018) \times 10^{-23}$WsecK^{-1}

10.1.1.6 Validation

The GIQE was validated using the data set tabulated in Table 10.2. The visible data set was split into two equal subsets. One was used for model development and the other for validation. The analysis on the development set resulted in (10.1). The value of R^2 was 0.986 and the standard error of prediction was 0.282 (NIIRS). When the equation was run on the validation data set, the R^2 was 0.934 and the standard error was 0.307. These values represent the estimated performance of the GIQE. Figure 10.4 shows results of the validation. Analysis of residuals for each of the terms in the equation showed no significant correlations between prediction errors and values of the variables.

In the case of the IR, the visible equation was applied to the IR data set to determine necessary adjustments. Aside from a 0.5 NIIRS adjustment to account for differences between the visible and IR NIIRS, no other adjustments were made. SNR values, however, did account for differences in wavelength and related phenomenology. A 2-degree K temperature difference between target and background temperature was used. Validation on the total IR data set showed an R^2 of 0.80 and a standard error of 0.38 (NIIRS). Figure 10.5 shows validation results. Residual analysis showed an R^2 value of 0.034 for the G/SNR term (which was statistically significant, but very weak). None of the other terms showed statistically significant relationships.

Table 10.2
GIQE Imagery Data Set

Parameter	Minimum	Maximum
GSD	3 in.	238 in.
RER	0.2	1.3
H	0.9	1.39
G	1	19
SNR	2	130

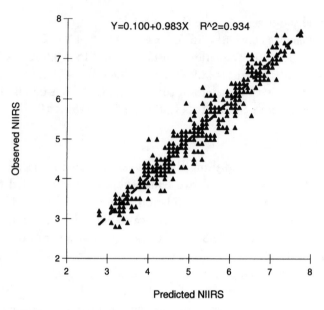

Figure 10.4 Validation results for visible data. (Data from [9].)

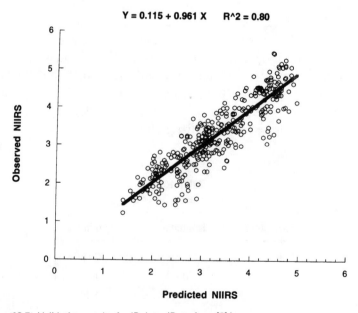

Figure 10.5 Validation results for IR data. (Data from [8].)

10.1.1.7 Limitations and Issues

The GIQE in its current form assumes optimized hardcopy viewing, no MTFC, and a $\lambda f N/p$ of 1. The $\lambda f N/p$ of a system is a measure of the sampling of the point spread function. A system with a $\lambda f N/p$ value of 2 will be sampled at the Nyquist frequency. Data from Chapter 8 can be used to estimate the effects of SC viewing as well as some BWC algorithms and rates. A recent study showed the effect of modifying $\lambda f N/p$ by increasing the along-scan sampling rate [12]. The cross-scan rate is fixed by the detector design. Increasing the along-scan rate effectively improves GSD, but also affects the system MTF (and thus RER) and SNR. Figure 10.6 shows results for sampling rates of 1.0× to 3.0×. Performance reached a maximum improvement of 0.35 NIIRS at a 2.0× rate. This compares to a predicted improvement of 0.5 NIIRS based on the GSD improvement alone.

In a second study [13], the effects of design variations at $\lambda f N/p$ of both 1.0 and 2.0 were studied. At a $\lambda f N/p$ of 1.0, sampling and aperture size were modified to provide a 2× change in relative GSD. At a $\lambda f N/p$ of 2.0, integration time and smear were changed and relative GSD was constant. Although the GIQE indicates that reducing GSD by a factor of 2 will provide a 1.0 NIIRS improvement, reduction by increasing $\lambda f N/p$ also affects MTF and SNR. It was shown that if no smear were present in the $\lambda f N/p = 1.0$

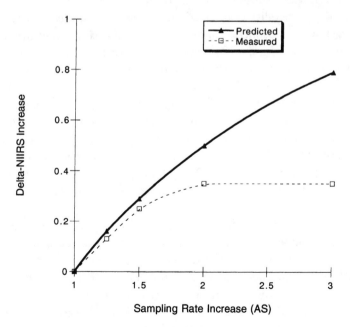

Figure 10.6 Effect of increased along-scan sampling. (Data from [12].)

design, the improvement at a $\lambda fN/p$ = 2.0 design would be ~0.6 NIIRS as opposed to the 1.0 NIIRS predicted due to a GSD improvement alone. The $\lambda fN/p$ was increased by decreasing GSD and increasing integration time to maintain the same signal level. The loss in MTF reduced the impact of the GSD change. With a one pixel smear in the $\lambda fN/p$ = 1.0 design, NIIRS degrades as $\lambda fN/p$ increased to 2.0.

The current version of the GIQE defines GSD in the ground plane. Version 3.0 defined GSD in the plane orthogonal to the line of sight. The change was made based on a regression comparison in the validation effort; use of the ground plane GSD provided a significantly higher correlation. The data used in the validation study had look angles as low as 25 degrees (angle between ground plane and line of sight from the target to the sensor). To date, no data has been available to define a look angle crossover point between the ground and the orthogonal plane. It is also the case that target orientation plays a more important role as look angle decreases.

10.1.2 Target Acquisition Sensor Performance Modeling

The target acquisition community has characterized imaging system performance with the minimum resolvable temperature (MRT) parameter, the minimum resolvable contrast (MRC) parameter, and an acquisition model that gives the probabilities of detection, recognition, or identification for a given target and atmospheric conditions. These probabilities of discrimination are plotted as a function of range and are used to determine the effective range of a target acquisition weapon system. MRT or MRC can be predicted with a sensor model, measured, or specified in a sensor design. The operational design requirement for a target acquisition sensor might be that it "must engage a T-72 tank with greater than a 90% probability of identification at 3 km in a standard U.S. atmosphere." An MRT or MRC requirement is derived from this operational requirement. Target acquisition sensors are found on M1 Abrams tanks, M2 Bradley fighting vehicles, AH-64 Apache helicopters, F-16 Falcon aircraft, UAVs, and a large number of other tactical platforms.

The "probabilities of discrimination" technique is a model that is dependent on a number of parameters. These include the target dimensions and the target-to-background contrast. The target-to-background contrast is described by an equivalent blackbody temperature differential for infrared (2–14 μm) imagers and by reflected light luminance differential for EO (0.4–2 μm) imagers. The atmospheric degradation of the contrast is responsible for an apparent differential temperature or apparent contrast at the imager entrance pupil. Imager performance is described by the MRT for infrared imagers and is described by MRC for EO imagers. These system performance descriptions are dependent on the sensitivity and resolution characteristics of the system, which, in turn, are dependent on a large number of parameters such as focal length, entrance pupil diameter, detector size, etc. The MRT and MRC are applied to the apparent target-to-background characteristics to give a frequency response; this can then be compared to Johnson's criteria to obtain a probability of detection, recognition, or identification. We first describe Johnson's criteria and the sensor performance parameters.

10.1.2.1 Johnson's Criteria

Researchers have demonstrated that, for a particular class of imaging sensors, the level of target discrimination is related to the sensor's limiting resolution. Static sensor performance is described as the case in which an observer has an infinite amount of time to perform the discrimination task on a non-moving target. It is now common practice to represent objects with bar targets for the purpose of sensor evaluation, analysis, and design. Johnson [14] conducted experiments at the U.S. Army's Night Vision and Electronic Sensors Directorate (NVESD) to develop requirements for different classes of sensors for detection, recognition, and identification of objects; in particular, tactical ground targets. Consider the targets in Figure 10.7 and their bar target equivalents. Johnson determined the number of bar pairs, or cycles, subtended by an object to allow performance of a discrimination task. Further NVESD experimentation yielded similar results, except that a length

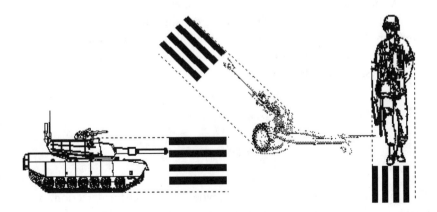

Figure 10.7 Objects and corresponding bar targets.

Table 10.3
Cycle Criteria

Task	Description	1-D cycles across minimum dimension	2-D cycles across characteristic dimension
Detection	Reasonable probability that blob is a tactical military vehicle	1.0	0.75
Recognition	Class discrimination (truck, tank, APC)	4.0	3.0
Identification	Object discrimination (M1A, T62, T72)	6.4	6.0

correction was introduced. The length correction improved the target signal by the square root of target length. The cycle criteria for the tasks of detection, recognition, and identification are described in Table 10.3. Note that the 1-D description corresponds to the number of cycles across the minimum dimension of the object. The 2-D cycle requirement is given for the characteristic target dimension (a function of target width and height as seen by the sensor):

$$d_c = \sqrt{W_{tgt} H_{tgt}} \quad \text{[meters]} \qquad (10.20)$$

A more accurate rendition of a characteristic dimension is the square root of the target area if the target were considered a silhouette. The cycle criteria in Table 10.3 correspond to probabilities of 50% for the discrimination task given. If the probability for a given number of cycles N across a target is desired, it is determined by

$$P(N) = \frac{\left(N/N_{50}\right)^{2.7+0.7(N/N_{50})}}{1 + \left(N/N_{50}\right)^{2.7+0.7(N/N_{50})}}. \qquad (10.21)$$

This equation is a "per task" equation where N_{50} is set for detection, recognition, or identification. For example, if the probability of recognition were desired as a function of the number of bar pairs across the target, N_{50} would be set to 3.0 and $P(N)$ would be plotted as a function of N. The 2-D cycle criteria are used more frequently than the 1-D criteria.

10.1.2.2 MRT Difference and MRC

Schade [15] is considered to be the founder of modern-day imaging system models based on his work in the 1950s and 1960s. Schade derived a performance measure for photographic, motion picture, and television systems [16]. The model was derived from an observer resolving a standard three-bar Air Force target in the presence of noise. Sendall, Rosell, and Genoud modified Schade's model for application to imaging IR systems where the resulting MRT measure applied to four-bar targets [15–17]. Barnard, Lawson, and Ratches further modified the thermal imaging model, resulting in the U.S. Army Night Vision Laboratory (NVL), now known as the Night Vision Electronics Sensors Directorate (NVESD), static performance model [17]. Many scientists and engineers have been involved in the refinement of the NVESD static model over the past two decades. This has produced the FLIR92 MRT sensor model [18] and the NVTherm sensor model [19]. The NVESD MRT theory was also applied to EO systems in terms of the MRC. The latest work has been in the area of improved noise characterization and modeling [20–23]. The noise-equivalent temperature difference (NETD) describes the sensitivity of an IR system, and the MTF describes the resolution of an IR system. The noise-equivalent input (NEI) and MTF give the sensitivity and resolution for EO systems. These functions are considered separable, but the actual sensitivity and resolution performance of an IR sensor is not separable. MRT and MRC describe sensitivity as a function of resolution and are considered the primary infrared and EO performance parameters.

MRT is defined as the differential temperature of a four-bar target at which the target is just resolvable by a person when using a particular sensor [24]. MRT is a sensor parameter that is a function and not just a value. It provides sensor sensitivity as a function of four-bar target frequency (i.e., resolution). The MRT is measured using 7:1 aspect ratio bars, as shown in Figure 10.8, resulting in a square four-bar target. The target-to-background differential temperature is varied until the four bars are just resolvable, and the differential temperature is increased from a small value until the bar target is just resolved. The differential temperature is plotted for each target frequency, where the collection of data points is the MRT. Note that the MRT curve appears to be related to the MTF curve in an inverted manner. This is indeed the case, as MTF is seen in the MRT model. While the modeling of MRT has taken different approaches, the measurement of MRT is considerably more standard.

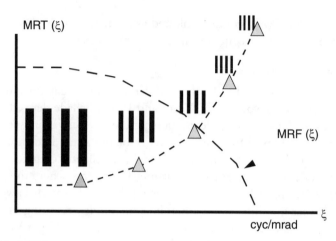

Figure 10.8 MRT Curve.

Analytically, the equation for the 1-D MRT is

$$MRT(\xi) = \left[\frac{1}{H(\xi)}\right]\frac{SNF \times FN^2\pi\xi\sqrt{B_W B_L}}{\delta\, F_o D^*_{\lambda-peak}\sqrt{2t_e}\,\eta_{eff}\,\tau S_L} \qquad (10.22)$$

where

SNF = visual threshold function (unitless)

FN = F-number of the optical system (unitless)

ξ = spatial frequency variable (cycles/milliradian)

$H(\xi)$ = system MTF (unitless)

F_o = effective focal length of the optical system (centimeters)

$D^*_{\lambda-peak}$ = peak (as a function of wavelength) specific detectivity of the detector(s) (centimeters times the square root of hertz per watt, or Jones)

t_e = eye integration time (seconds)

η_{eff} = scan efficiency of the sensor (unitless)

τ = equivalent in-band optical transmission (unitless)

η_{eff} = (fill factor) (actual dwell/available dwell); efficiency for staring sensor,

$\eta_{eff} = \eta_{scan}N_dA_{det}/\{(FOVH)[(FOVV)F_o^2]\}$; efficiency for scanning sensor, where η_{scan} is the scan efficiency, N_d is the total number of detectors, and A_{det} is the area of a detector.

There are also four integrals that must be evaluated: two spatial noise integrals, a spatial signal integral, and a detector response (sensitivity) integral:

$$B_W = W^2 \int_{-\infty}^{\infty} [H_N(\xi)H_e(\xi)H_W(\xi)]^2 d\xi \quad [\text{cm}^2]$$

$$B_L = L^2 \int_{-\infty}^{\infty} [H_N(\xi)H_e(\xi)H_L(\xi)]^2 d\xi \quad [\text{cm}^2]$$

$$\text{(10.23)}$$

$$S_L = L^2 \int_{-\infty}^{\infty} [H(\xi)H_e(\xi)H_L^2(\xi)]^2 d\xi \quad [\text{cm}^2]$$

$$\delta = \int_{\delta\lambda} \frac{\partial L(\lambda, T)}{\partial T} S(\lambda) d\lambda \quad [\text{W/cm}^2\text{-sr-Kelvin}]$$

where

W = bar target width (milliradians)
L = bar target length (milliradians)
$H_N(\xi)$ = transfer function operating on the noise (unitless)
$H_e(\xi)$ = human eye transfer function (unitless)
$H_W(\xi)$ = MTF of the bar pattern width (unitless)
$H_L(\xi)$ = MTF of the bar pattern length (unitless)
$H(\xi)$ = system MTF as stated (unitless)
$L(\lambda, T)$ = background radiance (watts per [centimeters squared times steradians times micrometers])
T = background temperature (degrees Kelvin)
$S(\lambda)$ = detector responsivity as a function of wavelength normalized to peak response (unitless).

A parameter very similar to MRT for EO systems is that of MRC. The concept is the same as MRT, but the four-bar response is for contrast targets seen by EO systems. MRC in the vertical direction is given by

$$\text{MRC} = \left[\frac{1}{H(\xi)} \right] \frac{\pi^2 SNF \times N_f \xi \sqrt{B_W B_L}}{16 S_L \sqrt{t_e E_{av}}} \quad \text{(10.24)}$$

where parameters are as already defined, except that N_f is a noise factor to account for less than ideal performance and E_{av} is the average electron flux from the cathode per steradian in number of electrons per steradian.

While the preceding expression is for tube cameras, there is a similar expression for CCD cameras. MRC is slightly different from MRT because the noise and background currents depend on the amount of light falling on the detector. The noise includes readout, shot, and fixed pattern noise terms. Shot noise is a function of target-to-background contrast, so MRC is a family of curves that depends on light level (i.e., an MRC curve is given for a particular light level).

A two-dimensional MRT or MRC curve is obtained with a combination of horizontal and vertical curves. The MRT, or MRC, values are matched in the horizontal and vertical directions and the corresponding spatial frequencies ξ_x and ξ_y are noted. The matched value is paired with a new two-dimensional spatial frequency

$$\rho_{2D} = \sqrt{\xi_x \xi_y} \tag{10.25}$$

The two-dimensional spatial frequency paired with the set of matching MRT or MRC values comprises the two-dimensional MRT or MRC.

10.1.2.3 Acquisition

The procedure for producing a probability of detection, recognition, or identification curve is quite simple. Consider the procedure flow as given in Figure 10.9. There are four parameters needed to generate a static probability of discrimination curve as a function of range: the estimated target-to-background temperature differential, an estimated height and width of a target, an atmospheric transmission estimate within the band of interest for a number of ranges around the ranges of interest, and the sensor two-dimensional MRT or MRC (either modeled or measured on a real system).

The target parameters are determined first. The characteristic dimension of the target is taken as the geometric mean of the target height and width. Recall that a more accurate characteristic of the target is to take the square root of the projected target area. The target-to-background temperature difference is then estimated based on target and background characteristics. For ground targets, these differential temperatures are usually between 1.25° and 4.0°C. The atmospheric transmission is then determined, and an equivalent blackbody apparent temperature is calculated based on the atmospheric signal reduction. The determination of atmospheric transmission $\tau(R)$ is not trivial and requires a sophisticated atmospheric model such as MODTRAN or LOWTRAN [10].

Once an apparent differential temperature is obtained, the highest corresponding spatial frequency that can be resolved by the sensor is determined. This is accomplished by finding the spatial frequency (on the MRT curve)

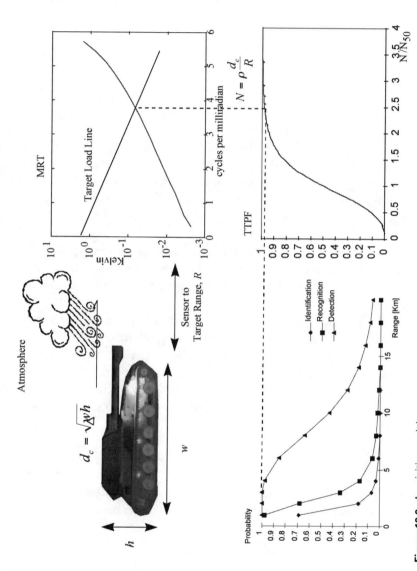

Figure 10.9 Acquisition model.

that matches the target-apparent differential temperature. The number of cycles across the characteristic target dimension that can actually be resolved by the sensor at a particular range then determines the probability of discriminating (detecting, recognizing, or identifying) the target at that range:

$$N = \rho \frac{d_c}{R} \tag{10.26}$$

where ρ is the maximum resolvable spatial frequency in cycles per milliradian, d_c is the characteristic target dimension in meters, and R is the range from the sensor to the target in kilometers.

The probability of discrimination is determined using the target transfer probability function (TTPF) given by (10.21). The level of discrimination (detection, recognition, or identification) is selected from Table 10.3 and the corresponding 50% cycle criteria N_{50} is taken from the table. The probability of detection, recognition, or identification is then determined with the TTPF for the number of cycles given by (10.26). The probability of discrimination task is then assigned to the particular range. A typical probability of discrimination curve will have the probability plotted as a function of range. Therefore, this procedure would be repeated for a number of different ranges.

While the following may be obvious, there are a number of characteristics that improve probability of detection, recognition, and identification in IR systems. Improvements are seen with larger targets, larger target-to-background differential temperatures, larger target emissivities, larger atmospheric transmission, smaller MRT values (as a function of spatial frequency), and usually small fields of view if the target does not have an extremely small differential temperature.

10.1.2.4 Examples

The MRT and MRC techniques were applied to two sensors, a wide field-of-view midwave IR sensor and a narrow field-of-view EO sensor. These sensors were selected because of the significant differences between their characteristics.

Midwave IR Example

The tactical model was applied to a midwave, wide field-of-view IR imager. A background temperature of 300 K was assumed in a sensor band of 3.5–6.0 μm. The sensor optics comprise a 1.5 f-number lens with a 10-cm effective focal length. The optical transmission was 0.8. The sensor frame rate was set to 30 Hz with two fields per frame. The detector was a 640 × 480 PtSi staring array with a 20 × 20 μm detector size. The detectors were spaced by 30 μm in each direction. The peak detectivity of the detectors was 5×10^{11} Jones. The display was a 10-fL CRT with a 15.24-cm display height and a 30-cm viewing distance. The eye integration time was set to 0.1 second and the eye SNR threshold was set to 2.5. The field of view of the sensor was 11 by 8.2 degrees. The NETD was determined to be 0.096°C, but this value was increased by a factor of 1.4 for the effect of nonuniformity to be 0.121°C. The combined MRT was calculated with the preceding parameters and the system horizontal and vertical MTFs. A few of the two-dimensional MRT values were 0.011, 0.018, 0.029, 0.047, 0.076, 0.125, and 0.203 K at spatial frequencies of 0.251, 0.403, 0.579, 0.799, 1.040, 1.277, and 1.493 cycles/mrad, respectively.

The acquisition model was applied to two targets to show the effects of target size. The first target was the standard NATO target 2.3 × 2.3 m in size at a target-to-background temperature of differential 1.25°C. The second was the top view of an M1 Abrams tank at 3.6 × 7.9 meters at a 1.25°C differential temperature. The atmospheric transmission was set at 0.85/km over the ranges that were used in the calculation. The discrimination criterion, N_{50}, was set at 3 cycles across the target for a 50% probability of recognition. The results of the acquisition calculations are shown in Figure 10.10.

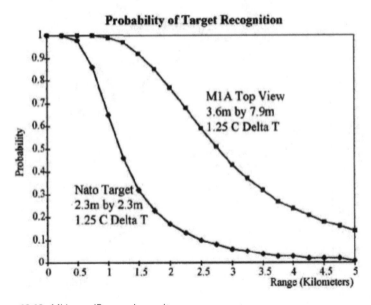

Figure 10.10 Midwave IR example results.

EO Example

Analyses were also performed on a narrow field of view EO system. The system field of view was 1.0 by 1.2 degrees with a spectral bandpass of 0.65–0.90 μm. The *f*-number of the sensor was 6.1 and the optical transmittance was 0.91. The focal length was 167 cm and the entrance pupil diameter was 11.3 cm. A 480×640 CCD detector array was used with 15- $\times$ 15-μm elements. The center-to-center spacing was 21.8 μrad for a 67% fill factor. The optics were not diffraction limited, and an estimate for the optical MTF was used. The display and eye characteristics were set as identical to those given in the IR example. The MRC of the system was calculated in both directions and then the two-dimensional MRC was determined. The MRC was calculated with a target-to-background contrast of 0.5 and an illumination of 5000 lux (bright daylight). This contrast was assigned to the NATO target and the top view of an M1A tank. A few values of the calculated MRCs were 0.0006, 0.0021, 0.0050, 0.0107, 0.0233, 0.0515, 0.1215, and 0.3515 (unitless) for spatial frequencies of 1.2, 3.6, 6.0, 8.4, 10.8, 13.2, 15.6, and 18 cycles/mrad. The acquisition results for the recognition task are shown in Figure 10.11.

10.1.2.5 Search

The equations and examples relating to the NVESD/PD model apply to a static situation—the target and the display are not moving and the target has

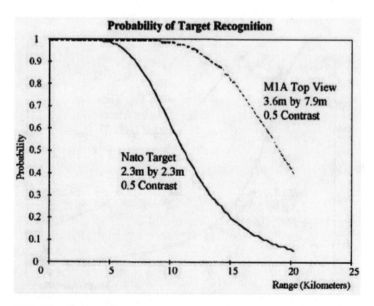

Figure 10.11 EO example results.

been fixated by the observer. A search term has been developed as a part of the NVESD Acquire model [25]. The term is empirically based and assumes that the target will be found with a probability approaching an asymptote at the probability of static recognition. The field of regard (the area to be searched) is divided into M fields of view, the field of view of the sensor. The probability of detection is defined as

$$Pd(t) = Pd_\infty(Pd_S) \qquad (10.27)$$

where $Pd(t)$ is the probability of detection in time t, Pd is the probability of detection as defined by (10.21), and Pd_S is the correction factor for search. It is defined as

$$Pd_S = \left[1 - e^{-t/M\tau}\right] \qquad (10.28)$$

where M is the number of FOVs in the FOR, and τ is a time constant defined as

$$1.7 \le \tau = \frac{6.8}{N/N_{50}} \qquad (10.29)$$

where 1.7 is a limiting time and N/N_{50} is as defined by the TPFF. In some circumstances, the limit may be disregarded [26]. The model assumes an average fixation time of 0.3 seconds and disregards interfixation times.

10.1.2.6 Validation

The MRC and MRT portions of the NVESD/PD model were developed in a series of studies and were validated by that work. Even earlier, Rosell and Willson [27] showed general agreement with the Johnson criteria but disagreed in terms of individual vehicles. Figure 10.12 shows the results of a study in which the TTPF model was used to predict the probability of tank identification. Twelve tanks at different aspect ratios (normalized for area) and 20 observers were used. Although the TTPF model overshoots at low blur and undershoots at large blur, the data still shows a strong correlation. The nonlinear form of the correlation is a result of the over- and undershoot. Figure 10.13 shows the same data with points for individual tank models shown. The spread results from the differences among the targets in terms of ease of identification. As noted in Chapter 7, some targets are more easily

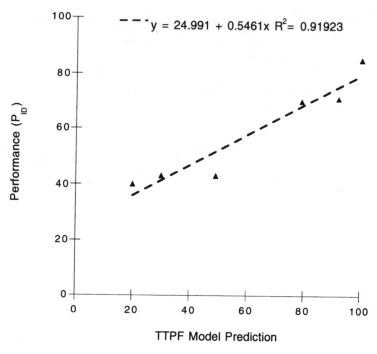

Figure 10.12 Comparison of TTPF predictions and observer performance for tank identification.

confused with other targets. In the case of the data set shown in Figure 10.13, the T-62 and T-72 were easily confused.

Validation of the search term has also been performed using data from several different field experiments [28]. Results were shown comparing field

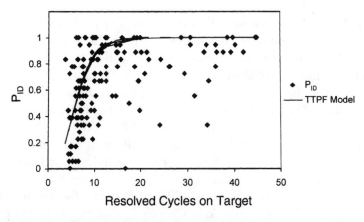

Figure 10.13 Effect of individual vehicle comparisons on TTPF agreement.

observations with model predictions. Probability of detection differences ranged from 0 to 0.2 with both over and under predictions of actual performance. Errors tended to approach 0 as search time increased, suggesting that the problem was with the search term as opposed to the TTPF.

10.1.2.7 Limitations

The Johnson criteria were developed for image intensifier applications. In that context, sampling and aliasing were not issues, and this is a limitation of the NVESD FLIR 92 model. Considerable effort is under way to resolve this limitation and a new model, NVTHERM has been developed to address sampling [19]. In a study comparing the effects of in-band aliasing edge shifts (line width variations) with out-of-band aliasing such as rastering, a task dependency was shown [29]. In-band aliasing affected tasks such as recognition (tested with symbol recognition) but had little effect on identification. The reverse was true for out-of-band aliasing. The spurious response function was defined as

$$SR = \frac{\int_{-\infty}^{\infty}(\text{spurious response})d\xi}{\int_{-\infty}^{\infty}(\text{baseband MTF})d\xi}$$

$$SR_{in\text{-}band} = \frac{\int_{-v/2}^{v/2}(\text{spurious response})d\xi}{\int_{-\infty}^{\infty}(\text{baseband MTF})d\xi}$$

$$SR_{out\text{-}of\text{-}band} = SR - SR_{in\text{-}band} \tag{10.30}$$

where v = sample frequency (2π times cycles per milliradian) and ξ = spatial frequency (2π times cycles per milliradian). The effect of spurious responses was captured by a reduction or "squeeze"of the system MTF. The MTF squeeze was defined as a contraction of the MTF frequency axis in both the horizontal and vertical dimension and was applied to the signal MTF. The MTF squeeze for recognition was defined as

$$MTF_{squeeze} = (1.0 - 0.32 SR_H)^{1/2}(1.0 - 0.32 SR_V)^{1/2} \tag{10.31}$$

For identification, the MTF squeeze was defined as

$$MTF_{squeeze} = (1 - 2 SR_{H\text{-}out\text{-}of\text{-}band})^{1/2}(1 - 2 SR_{V\text{-}out\text{-}of\text{-}band})^{1/2}. \tag{10.32}$$

A baseline probability curve was developed to define the relationship between probability of recognition and the MTF cutoff using the TTPF. It had the form

$$P_R = \frac{(MTF_{cutoff}/K)^E}{1+(MTF_{cutoff}/K)^E} \qquad (10.33)$$

where

$$E = 2.7 + 0.7\left(\frac{MTF_{cutoff}}{K}\right)$$

and K was defined by a fit to the data.

A somewhat different approach was taken to the problem of sampling in the laboratory by developing laboratory measurement procedures to account for sampling effects [30]. Procedures for the measurement of sampled imagery were also developed for measurement of 3-D noise (IETD), MTF, and MRT. Further discussion of the treatment of sampling effects is provided in [31].

10.1.3 Physique

Despite a lack of published validation, Physique [3] is of interest because it is based in information theory and does not require imagery to be run.[3] The model is defined as

$$I = \pi \int_0^{\xi_c} \xi \log_2(S/N + 1)d\xi \qquad (10.34)$$

where

I = information content
ξ = ground spatial frequency
ξ_c = system cutoff frequency
S = signal power spectrum weighted by MTF^2
N = noise power spectrum weighted by MTF^2.

---

3. The reference document indicates that the model was validated against over 20,000 EO ratings on an IA scale and that "Physique performed well." No data is provided on the actual results.

Information can, in turn, be related to performance measures such as NIIRS or probability of recognition. Physique computes information in the along- and cross-scan directions separately and then uses the geometric mean of the two values. Signal and noise are assumed to be Gaussian, stationary, ergodic, statistically independent, and mean-square additive. Physique assumes a hardcopy (film) output.

A diagram of the model is shown in Figure 10.14. Target geometry (block 1) includes target type, orientation, height, density, and reflectance.

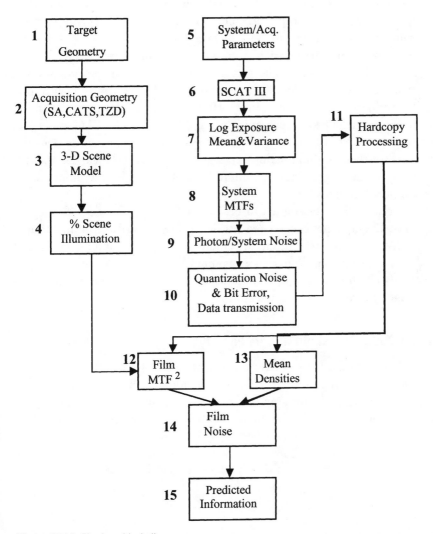

Figure 10.14 Physique block diagram.

Acquisition geometry (block 2) includes solar altitude, sensor-target-sun angle, and target zenith distance. A 3-D scene model (block 3) is used to calculate the percentage of the scene illuminated by each of five illumination types: daylight vertical, daylight horizontal, shadow vertical front-lit, shadow vertical back-lit, and shadow horizontal (block 4). The information content is computed separately for each of the five types, and a weighted average is then computed. Block 5 represents system and acquisition parameters. These include such things as f-number, focal length, filtration, solar altitude, vehicle altitude, atmospheric transmittance, obliquity, and aspect angle of the sun and look vectors. Target reflectance mean and variance are also included.

In block 6, an atmospheric model called SCAT III [32] computes atmospheric transmittance, target irradiances, and sky radiance. Block 7 computes exposure mean and variance, and block 8 computes system MTFs due to optics, smear, cross-talk, and focus. In block 9, photon and system noise are computed. Quantization noise, bit errors, and data transmission encoding and decoding are calculated in block 10. Block 11 handles all of the effects involved in writing to hardcopy including MTFC, DRA, TTC, as well as writing device MTFs and film sensitometry. Film MTF and mean densities for each of the five illumination types are computed in blocks 12 and 13. Film noise due to graininess is computed in block 14. Finally, information is computed in block 15.

Conceptually, Physique can be used for visible, IR, and MS imaging systems. Physique has been coded and is reportedly available on mainframe and IBM PC computers. Inputs are entered in 11 groups and total over 200 pieces of information. From that viewpoint, Physique appears to be the most explicitly all-encompassing prediction model. The lack of published validation data, however, limits applications.

10.1.4 Radar Threshold Quality Factor

The RTQF model [5] was developed in the early 1970s.[4] The RTQF included system design–related parameters (impulse response, sidelobe levels, noncoherent integration, amplitude transfer), a visual system model, and display geometry. The RTQF model defines a value of quality, called Q, as

$$Q = \int_0^\infty \int |H_I(u,v)| \, dudv - \int_0^\infty \int C(u,v) \, dudv \qquad (10.35)$$

---

4. The RTQF is © 2000 Veridian Systems/Veridian ERIM International Inc.

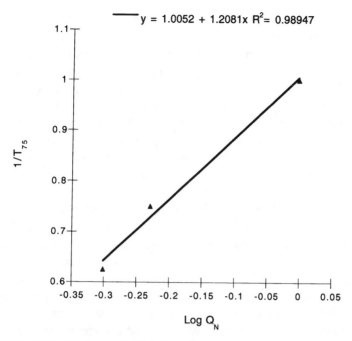

Figure 10.15 RTQF validation results. (Data from [5].)

where $|H_I(u,v)|$ is the two-dimensional system MTF and $Z = \int_0^\infty \int C(u,v)dudv$ represents the portion of the image at spatial frequencies not of use to the observer. The form of $C(u,v)$ is related to the human contrast sensitivity function.

The RTQF explicitly includes the system impulse response function, clutter (additive), system noise, and noncoherent integration. A thresholding approach to dealing with peak sidelobe levels, integrated sidelobe levels, and geometric fidelity, as a function of task was proposed, but explicit thresholds were not defined.

A small validation study was performed using SAR data at 10–20 ft IPR. Two-sample noncoherent integration was also included in the data set, and targets were distributed strategic complexes (total of 18). Nine experienced radar operators participated in the evaluation. Results were scored in terms of the reciprocal of the time required to achieve a 75% probability of correct recognition ($1/T_{75}$). Figure 10.15 summarizes the results.

10.1.5 Other SAR Models

Other than the RTQF, no models predict the performance of human observers for SAR. The emphasis on SAR has been on modeling either the

physical image quality characteristics of imagery or the performance of automatic target detection/classification systems. SAR imagery has properties
quite different from visible and IR imagery, and basic human perception studies do not relate well to an understanding of SAR recognition and identification. Because the SAR itself is the energy source and the effects of the
atmosphere and weather can largely be ignored, the problem of machine recognition may be considered more tractable (although this is not necessarily
the case). Whatever the reasons, there appear to be no current models to
predict target recognition and identification for high-resolution SAR systems.

Given an SAR system design, models are available to predict the image
quality characteristic of the imagery produced by the system. Such a model
was used in the development of the RTQF. Image quality is defined in terms
of the system point spread function or impulse response, noise, and clutter.
Extensive work has also been performed to model and define the relationship between radar cross-section (RCS) and image intensity [33]. Target
detection entails making a distinction between a target return and background
clutter and noise. At the single pixel level, decision rules are applied to separate targets from false alarms based on intensity (or RCS) differences. As resolution improves, targets are defined by multiple scattering centers, and the
complexity of the problem increases. The ability to classify and identify targets on the basis of the location and strength of these scattering centers is
achieved for military targets (e.g., vehicles) as the resolution (−3 dB IPR)
improves to on the order of 1 ft. The appearance of these targets, however,
varies substantially as a function of viewing geometry. Thus, a given target
may have several thousand patterns of returns or signatures as a function of
imaging geometry. One approach to this problem is to generate synthetic
images of the target based on knowledge of scattering properties. These can
then be matched to real images. Other approaches extract signatures from
real SAR data and use those in a pattern-matching process to classify and
identify targets [34]. Probabilities of 0.9 and higher are reported for tank
identification on Moving and Static Target Acquisition and Recognition
(MSTAR) imagery at 1 ft IPR with what appears to be a relatively benign
background.

10.2 Image-Based Models

As the name implies, image-based models require the availability of imagery. The image quality model (IQM), an information-based metric, predicts
NIIRS from single images [35,36]. The Sarnoff JND model initially predicts
perceptual differences between two images [37,38]. Those differences, ex-

pressed in terms of the number of JNDs, have been related to both NIIRS ratings and subjective video quality ratings (DSCQS). Several similar models have been developed in an attempt to predict DSCQS ratings and are briefly reviewed. Thus far, none has been any more successful than the simple PSNR metric.

10.2.1 Image Quality Model

The IQM measures the power spectrum of a digital image and, with some adjustments, computes the information content of the image, which in turn is related to NIIRS. The adjustments include incorporating the visual system MTF, a noise adjustment, and a scale factor. The IQM is defined as

$$IQM = \frac{1}{M^2} \sum_{\theta=-180°}^{180°} \sum_{\rho=0.01}^{0.5} S(\theta_1)W(\rho)A^2(T\rho)P(\rho,\theta) \qquad (10.36)$$

where M^2 is the number of pixels in the image (for an image of $M \times M$ pixels), $S(\theta)$ is the scale factor, $W(\rho)$ is the modified Weiner noise filter, $A^2(T\rho)$ is the square of the MTF of the human visual system (HVS), and $P(\rho,\theta)$ is the two-dimensional power spectrum.

The scale factor is defined as the ratio of pixel pitch to ground distance defined in terms of cycles per ground meter. It is defined as

$$S\theta_1 = \frac{f}{2Dq} \qquad (10.37)$$

where f is the focal length, D is the distance between the sensor and the ground, and q is the pixel pitch.

The modified Weiner noise filter is computed as

$$W(\rho) = \left[\frac{2\pi a \sigma_s^2 \exp(-\rho^2/\sigma_g^2)}{2\pi a \sigma_s^2 \exp(-\rho^2/\sigma_g^2) + \kappa_1(a^2 + \rho^2)^{1.5}|N(\rho)|^2} \right]^{\kappa_2} \qquad (10.38)$$

where a is the reciprocal of the average pulse width, σ_s^2 is scene variance in bits/pixel, ρ is radial spatial frequency in cycles/pixel width, σ_g^2 is the variance of a Gaussian MTF with 20% modulation at Nyquist, π is the noise power spectrum, and κ_1 and κ_2 are empirically derived constants of 51.2 and 1.5, respectively, for noisy images, and 19.2 and 1.5 for a noise-filtered image.

The HVS MTF is defined as

$$A(T\rho) = (0.2 + 0.45T\rho)\ \exp(-0.18T\rho) \qquad (10.39)$$

where T is a constant defined as the spatial frequency of the peak of the HVS MTF (5.11 cycles/degree), p is cycles/pixel width, and $T\rho$ is cycles/degree subtended by the eye. The MTF peak location was set at 20% of the 0.5-cycle/pixel width display Nyquist frequency such that T was set at 51.1.

Finally, the normalized two dimensional power spectrum was defined as

$$P(\rho,\theta) = \frac{\left|H(\rho,\theta)^2\right|}{u^2 M^2}\ \text{and}\ \theta = \tan^{-1}\frac{v}{u} \qquad (10.40)$$

where $|H(\rho,\theta)|^2$ is the two-dimensional power spectrum (in polar coordinates), u^2 is the square of the average gray level of the image (dc power), M^2 is the number of pixels in the image, and v and u are the spatial frequency components of the input image. Because the two-dimensional power spectrum is radially symmetric, a one-dimensional spectrum can be created by averaging the energy in radial bands corresponding to frequency. Power is aggregated in bands that are 1/64 cycle/pixel width. Figure 10.16 shows an example of a one-dimensional power spectrum computed on a NIIRS 5 image.

The IQM requires as input a digital image. The IQM thus generally ignores display issues. There is no reason, however, why a properly controlled digital image of the displayed image could not be used as input. A two-megapixel digital camera could be used to capture a SC image display. Alternatively, the relationship between IQM values and NIIRS ratings could be established on a defined and well-characterized display and display environment. Hardcopy imagery can simply be digitized.

Although the power spectrum is theoretically independent of scene content, lower values will be obtained in empty areas (e.g., water). Consequently, the power spectrum should be computed on an area in the scene having cultural content. This same requirement exists for making NIIRS ratings.

The relationship between IQM and NIIRS was originally defined in a study using digitized HC imagery. Four analysts provided NIIRS ratings on

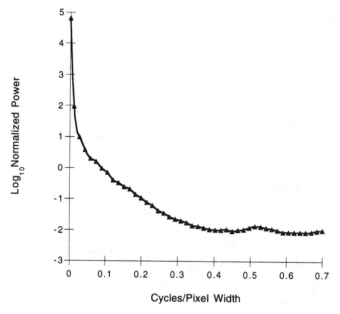

Figure 10.16 One-dimensional power spectrum.

50 digitized aerial photos displayed on a CRT. The relationship was originally defined as

$$\text{NIIRS} = 1.93 \log_{10} IQM + 8.77 \tag{10.41}$$

The R^2 value was 0.81.

The IQM is currently available in operating form at http://www.mitre.org/centers/cafc3/mtf/mainpage.html. It requires as input a raw image. The coefficients to predict NIIRS have been modified to predict NIIRS on the image "as is" or to predict it after haze removal (dynamic range adjustment). The two current equations are

$$\text{NIIRS} = 1.6092 \log_{10} IQM + 8.6849 \tag{10.42}$$

with haze, and

$$\text{NIIRS} = 2.2933 \log_{10} IQM + 8.0 \tag{10.43}$$

after haze removal. The current model also measures the direction and magnitude of one-dimensional smear. The power in radial wedges (0.5 degrees wide over 180 degrees) is computed and the wedge with the highest power defined. The ratio of power in that wedge to the power in the lowest power wedge at or near a right angle (to the highest power wedge) is computed and the ratio compared to a threshold value. If the value exceeds the threshold, the presence of smear is declared.

In its current form, the IQM predicts only visible NIIRS. Conceptually, it could presumably predict IR NIIRS, as well as the effects of bandwidth compression. Application to SAR would require additional development.

10.2.2 Sarnoff JND Model

The Sarnoff JND model computes the differences between two images in terms of the response of the human visual system expressed as just-notice-able-differences (JNDs).[5] The number of JNDs is then related to some measure of relative quality (e.g., DSCQS) or interpretability (NIIRS). The primary emphasis in recent years has been on the prediction of video quality (DSCQS) ratings.

The Sarnoff model builds upon several previous models. It directly follows the model developed by Carlson and Cohen [39]. In this model, the image one-dimensional power spectrum is decomposed into several discrete frequency bands and the output of each band subject to a static nonlinearity. The nonlinearity is compressive for large input values and expansive for small. This model was shown to predict changes in edge sharpness, for example. The square root integral (SQRI) model developed by Barten [40] is similar but is easier to compute. Here, the separate discrete frequency bands are replaced by a single integral over frequency bands. Other similar 1-D models have been developed and the one-dimensional case has been extended to two dimensions [41]. Because of a desire to operate in real or near real time, the Sarnoff model attempts to blend speed and accuracy.

The Sarnoff model is diagrammed in Figure 10.17. Input to the model is a pair of digital images differing in some process (e.g., BWC, smear, etc.). In addition to the two images, four other input values are required. They are the image sample spacing, observer-to-image distance, visual fixation depth, and eccentricity of the image in the observer's visual field. Distances are measured in millimeters, eccentricity in degrees. The images are first convolved with an expression representing the PSF of the eye:

---

5. The original papers on the Sarnoff model identified it as the Visual Display Model (VDM). Other papers refer to it as the Sarnoff model. A commercialized version is called the JNDmetrix.

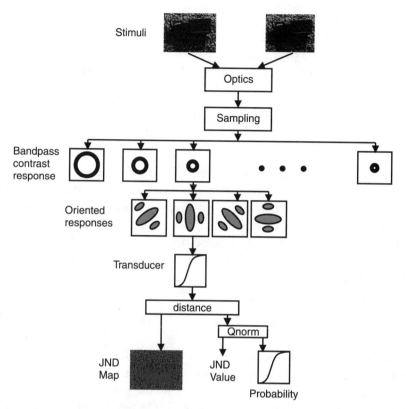

Stimuli

Optics

Sampling

Bandpass contrast response

Oriented responses

Transducer

distance

Qnorm

JND Map

JND Value

Probability

Figure 10.17 Sarnoff JND Model.

$$Q(\rho) = 0.952 \ \exp(-2.59 \ |\rho|^{1.36}) + 0.048 \ \exp(-2.43 \ |\rho|^{1.74}) \quad (10.44)$$

where $Q(\rho)$ is the intensity of light at a distance ρ relative to the maximum, and ρ is the distance in minutes of arc from a point of light.

At the sampling stage, the sampling of the image by the retinal cones is simulated. A Gaussian convolution and point sampling sequence results in (for foveal viewing) a retinal image of 512×512 pixels at a density of 120 pixels per degree of visual angle. For off-axis, nonfoveal viewing (eccentricity of greater than 0) the retinal pixel density is defined by

$$d = 120/(1 + ke) \quad (10.45)$$

where d is the pixel density, e is the eccentricity in degrees, and k is a constant set at 0.4.

The raw luminance signal is next converted to units of local contrast by decomposing the image into a Laplacian pyramid. This results in seven bandpass levels with peak frequencies of 32–0.5 cycles/degree with levels separated by one octave. In the continuous domain, this is expressed as

$$\hat{c}(\vec{x}) = \frac{I(\vec{x}) * (G_k(\vec{x}) - G_{k+1}(\vec{x}))}{I(\vec{x}) * G_{k+2}(\vec{x})} \qquad (10.46)$$

where $\hat{c}(\vec{x})$ is the contrast at pyramid level k, $(\vec{x})$ is a two-dimensional position vector, $I(\vec{x})$ is the input image, and $G_k(\vec{x})$ is a Gaussian convolution kernel.

Each pyramid level is next convolved with four pairs of spatially oriented filters consisting of a directional second derivative of a Gaussian and its Hilbert transform for each of four different orientations. The pair of Hilbert output values at each point are squared and summed to define a local energy measure

$$e_{k,\theta}(\vec{x}) = (o_{k,\theta}(\vec{x}))^2 + (h_{k,\theta}(\vec{x}))^2 \qquad (10.47)$$

where o and h are the oriented operator and its Hilbert transform, k is pyramid level, and θ indexes over the four orientations.

Each energy measure is normalized by the grating contrast detection threshold for the position and pyramid level. The threshold is calculated from Barten's square root integral model [40]:

$$1/M_t(v) = av\ \exp(-bv)\sqrt{1 + c\ \exp(bv)} \qquad (10.48)$$

where

$$a = \frac{540(1 + 0.7/L)^{-0.2}}{1 + 12/w(1 + v/3)^2}$$

$$b = 0.3(1 + 100/L)^{0.15}$$

$$c = 0.6$$

and L is display luminance in cd/m^2 and w is display width in degrees. The result of the normalization process is

$$\hat{e}_{k,\theta}(\vec{x}) = \frac{e_{k,\theta}(\vec{x})}{(M_t(v_k, L_k(\vec{x})))^2} \qquad (10.49)$$

where v_k is the peak frequency for pyramid level k, and L is the local luminance as used in (10.43).

At the transducer stage, each normalized energy measure is put through a sigmoid nonlinearity to reproduce the dipper shape of the human contrast discrimination function. The function is

$$T(\hat{e}_{k,\theta}(\vec{x})) = \frac{2(\hat{e}_{k,\theta}(\vec{x}))^{n/2}}{(\hat{e}_{k,\theta}(\vec{x}))^{(n-w)/2} + 1} \qquad (10.50)$$

where n, k, and w are constants defined as 1.5, 0.1, and 0.068, respectively [35].

A pooling stage convolves the transducer output with a disc-shaped kernel of diameter 5.

At this point in the process, the model output for each position in the image can be considered as an m-dimensional vector, where m is the number of pyramid levels times the number of orientations. In the operation called "distance," the distance between the vectors for the two input images is calculated. The smaller pyramid levels are first upsampled to the full 512×512 size. This produces for each image a set of m arrays (seven pyramid levels $\times$ four orientations = 28), where each entry is a transducer output.

The distance is then calculated as

$$D(\vec{x}_1, \vec{x}_2) = \left\{ \sum_{i=1}^{m} [P_i(\vec{x}) - P_i(\vec{x}_2)]^Q \right\}^{1/Q} \qquad (10.51)$$

where x_1 and x_2 are the two input images and Q is a parameter set at 2.4. This results in an output image where brightness is related to distance (number of JNDs). More commonly, the average distance is used, although the maximum can also be used. The average value (a JND value) can then be related to some measure of quality or interpretability.

10.2.2.1 Validation

The JND model was validated as a predictor of NIIRS difference ratings using a set of data from Snyder [42]. Four scenes were degraded with a matrix of

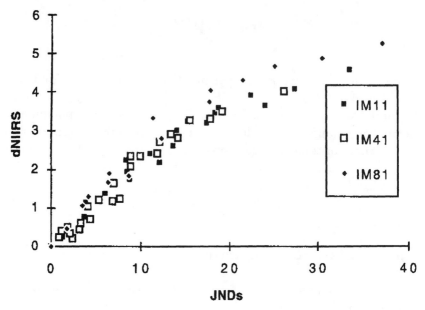

Figure 10.18   JND versus NIIRS.

five levels of blur and five levels of noise. IAs rated the NIIRS difference (delta-NIIRS) between the original and degraded versions of the images. A copy of the plot showing delta-NIIRS versus computed JNDs for three different input scenes is shown in Figure 10.18. No statistics were provided. Inspection of the plot suggests a nonlinear fit to the data with a standard error of less than 0.5 NIIRS.

A recent study [43] conducted by the Video Quality Experts Group (VQEG) compared the ability of several video quality measures in terms of their ability to predict DSCQS ratings on a variety of video scenes and hypothetical reference circuits (HRCs), primarily bandwidth compression. Two groups of eight HRCs (defined as low and high quality), 10 scenes, and two video formats (50 and 60 Hz) were used. Results were analyzed in terms of the correlation between model predictions and DSCQS ratings made in different labs. Not all images were seen by all raters. The JND model showed a correlation of 0.63 over the full data set, which was lower than, but not different statistically from, that shown by the PSNR metric. A plot of the data suggests a nonlinearity between the JND values and the DSCQS ratings. It should be noted that the correlations were computed on the scene by HRC matrix rather than data being averaged across scenes. Averaging across scenes (or targets) is the common procedure in reporting data from S&R studies—

this tends to increase the correlation. Therefore, the correlation of 0.63 reported for the Sarnoff metric cannot be directly compared to, for example, NIIRS correlations reported for other metrics.

10.2.2.2 Limitations

Although the Sarnoff metric has been reported as the de facto standard for objective video quality metrics [44], it has not yet been shown to outperform other metrics such as PSNR and RMSE [43,44]. Conversely, however, mean square error metrics do not always perform well [45]. One reported limitation of the Sarnoff model is the assumption of channel independence. Recent data suggests interaction between different cortex bands. A comparison of a within-channel masking transducer of a type similar to that used in the Sarnoff model was made with a between-channel transducer [45]. The correlation of a subjective rating with objective predictions was 0.87 for the between-channel version and 0.85 for the within-channel version.

The JND model is under continuous development. The commercialized JNDmetrix includes provision for color and temporal modeling. The temporal portion is currently being updated (personal communication, Dr. Jeffrey Lubin, April 2000).

10.2.3 Other Image-Based Models

Several other image difference models have been proposed and evaluated. In general, they attempt to model the response of the human visual system to complex stimuli largely using data from relatively simple stimuli studies. They may be further constrained to be computationally simple so that they can be applied in real time to a video data stream. Such models tend to work well when the image set is varied along a single continuum and work less well when varied across two or more dimensions or measures of physical quality. Virtually all of these models attempt to predict perceptual difference or quality ratings. Just as physical image quality ratings and measures of interpretability do not always correlate in a straightforward fashion, subjective quality ratings and interpretability may respond differently to image differences. The visible NIIRS is almost totally spatially based. Consequently, color deficiencies resulting from compression would be expected to have little effect. Such deficiencies might, however, impact subjective quality ratings.

In the previously referenced VQEG study [43], eight models in addition to the Sarnoff were evaluated. With one exception, they were all difference models. The exception (Centro de Pesquisa e Desenvolvimento) computes a set of objective parameters for an image segmented in terms of plane, edge, and texture regions. These objective parameters are related to subjective quality ratings in a logistic regression model. The model defines an estimated impairment level for each parameter; these estimates are combined for an overall value. Three of the difference metrics showed statistically significant lower performance than the PSNR metric, and there were no statistically significant differences among the remainder. There, thus, appear to be six models that predict DSCQS ratings with about the same accuracy as PSNR (which accounted for 56% of the variance in DSCQS ratings).

An interesting variation on the IQM has recently been evaluated [46]. Called a digital ring-wedge detector, the approach evolved from an analog optical power spectrum classifier. Digital ring-wedge data was provided to a neural net for training purposes, and the net was then tested with the remainder of the data set. In addition to the ring-wedge data on the original image FFT, an edge profile image was generated and an FFT of the edge profile run through the detector. Image sets included a blur set and a JPEG compression set, each at five quality levels. Classification accuracies of up to 98% were reported with knowledge of the original image.

Finally, we discuss an image difference model developed to deal with the issues of sampling and aliasing for the NVL model. This empirical model of recognition is of some interest because of the manner in which it deals with target identification [47]. The goal of the model is to deal with the effects of aliasing and sampling that occur with focal plane arrays. A series of evaluations were performed in which observers attempted to distinguish between and among various geometric shapes. A metric called recognition contrast was initially defined as follows:

$$C_R = \frac{L_{D_{\max}} - L_{D_{\min}}}{L_{\max} - L_{\min}} \qquad (10.52)$$

where $L_{D_{\max}}$ and $L_{D_{\min}}$ are the maximum and minimum irradiance differences between two superimposed images and L_{max} and L_{min} are the maximum and minimum irradiances in the target image. This metric was shown to have a nonlinear fit to the probability of recognition data. Targets were degraded in terms of Gaussian blurring and sampling and recognition performance compared to the values of C_R. Again, a nonlinear fit was shown. The results were found to be equivalent to the Johnson and NVL model data for detection and recognition. A fixed-pattern noise term, noise-to-signal ratio (NSR), was added using the relationship

$$NSR = \frac{\sigma}{(L_{max} - L_{min})} \tag{10.53}$$

where L_{max} and L_{min} are the maximum and minimum irradiances in the image, and σ is the standard deviation of noise defined by

$$\sigma = \sqrt{\sum \frac{(x_i - x_{avg})^2}{n}} \tag{10.54}$$

where x_i is the intensity at an image point, x_{avg} is the mean noise level, and n is the number of image points.

The noise and contrast terms were combined as follows:

$$C_R = \frac{L_{D_{max}} - L_{D_{min}}}{(L_{max} - L_{min})(1 + 25NSR)} \tag{10.55}$$

where terms are as defined in (10.52) and (10.53). A function was fit to the data of the form

$$P_T = \frac{1}{1 + 10^{-3.8} C_R^{-3.1}} \tag{10.56}$$

where P_T is the probability of recognition and C_R is as defined in (10.55). A plot of the data, however, indicates the fit is essentially linear over the range of $0 < C_R < 0.1$. Over this range, P_T goes from 0 to 0.98 (estimated). For identification, it was proposed that a difference image be defined by overlaying an image of the target and the object most likely to be confused with the target. Combined with the NSR value for the sensor, the probability of identification can then be computed. This approach, however, assumes that all identification cues are defined by the outline of the object. For battlefield target acquisition and for EO systems, this may be a reasonable assumption. For aerial imagery and for IR systems, it generally is not.

With the possible exception of Physique, none of the models discussed predicts the performance of multi- and hyperspectral systems. The MS IIRS can be defined in terms of published GSD data [48]. The equation for the MS IIRS relationship was defined as

$$MS\ IIRS = 9.18 - 2.4 \log_{10} GSD + 0.44 SWIR \tag{10.57}$$

where GSD is measured in inches[6] and SWIR indicates the presence of a SWIR band (SWIR = 1).

Knowledge of GSD and whether or not a SWIR band is present is sufficient to predict MS IIRS to an accuracy of 0.43 (standard error of prediction). It is also conceptually possible to use the GIQE to derive an approximate estimate of MS IIRS.

In the case of hyperspectral systems, image quality metrics have been developed and evaluated and a conceptual plan developed for the assessment of information content [49]. The image quality metrics represent an extension of the GIQE to the hyperspectral case. The metrics included spatial, spectral, and radiometric measures. The spatial metrics included GSD, RER, and geolocation accuracy. GSD and RER were defined as in the GIQE; geolocation accuracy is a measure of the accuracy with which image pixels can be related to their true ground position.

Spectral measures included measures of spectral resolution, spectral edge sharpness, system noise, and spectral calibration accuracy. Spectral resolution was defined as the full width at half maximum (FWHM) of the system spectral response (analogous to the –3 dB IPR for radar). Atmospheric data is used to measure the FWHM, and because the measure is affected by the accuracy of spectral calibration, a spectral calibration error is defined at the same time. Spectral resolution is analogous to spatial resolution. Spectral RER is defined from the spectral MTF and is analogous to spatial RER. System noise is defined in terms of noise equivalent delta radiance (NEΔL) and is defined as the input radiance that provides an SNR of 1.0. It is equivalent to dark noise, which is defined as

$$N_D = \sqrt{N_T^2 - S} \qquad (10.58)$$

where N_D is the dark noise, and N_T is the total noise at signal S. A Poisson noise distribution is assumed and units are photoelectrons. NEΔL is measured from two uniform regions in the data as

$$NE\Delta L = \sqrt{\frac{S_1 N_{T2}^2 - S_2 N_{T1}^2}{S_1 - S_2}} \qquad (10.59)$$

where N_T and S are defined as before.

---

6. The original equation defined GSD in meters.

Finally, radiometric accuracy was defined as the error between an expected (as measured in a laboratory) spectrum and a measured (by the sensor) spectrum. The truth spectrum is resampled to match the spectral response of the sensor, and the percent error is calculated for each wavelength in the spectrum. With the exception of NEΔL, measurements made on spatially and spectrally degraded HSI cubes were shown to produce predictable results [49]. It was concluded that in-scene noise measurements were inadequate and that dark noise measurements and calibration data are required to adequately measure NEΔL.

In a second study, two HSI sensors were evaluated using the spatial and spectral quality metrics [50]. With the exception of the spectral calibration metric, the measures were shown to produce expected results.

Spectra produced by HSI systems are used to identify materials on the ground. The measured spectra are compared to a catalog of known spectra. Both the measured and catalog spectra show variability due to variability in the material, more than one material in a measured pixel, directional dependencies of the spectra, calibration and measurement error, and noise. The process can be likened to distribution testing, the null hypotheses being one of no difference between the measured and cataloged spectrum. Conceptually, the probability of a correct identification can be defined as

$$P_I = f(SD), (SP), (AP), (SA), \beta \qquad (10.60)$$

where P_I is the probability of a correct identification, SD is the accuracy of spectrum definition, SP is the sensor performance (spatial, spectral, radiometric), AP is the performance of the analysis system (measurement and processing), SA is the sample abundance, and β is the decision criterion in the context of signal detection theory. The probability of a correct identification is a function of the uniqueness of the unknown spectrum relative to other incorrect spectra and the decision criterion (β) used in the processing algorithm. When a very close match is required, the probability of identification and the probability of incorrect identification decrease. When the decision criterion is relaxed, the reverse occurs. It was envisioned that for assessment of a single system, probability of correct identification on a sample of materials would be assessed for one or more levels of β. Where the objective was to compare two systems, differences in d' would be assessed.

10.3 Summary

Two classes of sensor performance models were presented. The first predicts some measure of performance from knowledge of sensor design and operating parameters. The GIQE is the current standard for S&R visible and IR systems; no comparable model exists for SAR. The GIQE is potentially extendable to MS systems but has not been evaluated in that context. The NVESD/FLIR 92 and NVTHERM models are the current standard for EO and IR targeting and acquisition sensors. Again, no comparable model exists for SAR or MS systems.

Physique is perhaps the most comprehensive of the parameter-based models but unfortunately, validation data has not been published. The RTQF is the only currently available model that predicts human observer performance for SAR imagery, but it has not been extended to the current generation of high-resolution SAR systems. A variety of models exist to predict the characteristics of an SAR image and to predict ATR performance.

All of the parameter-based models can also be used with imagery. The predominant image-based models are the power spectrum-based IQM and the Sarnoff JND model. Both have been shown to predict NIIRS for downward-looking visible systems. They are both theoretically extendable to IR and SAR (as well as MS), but performance in those applications has not yet been demonstrated. A variety of other video models have been developed, primarily for use in the commercial broadcast industry. Finally, a concept for the image-based modeling of HSI systems has been developed and an initial assessment of image quality metrics performed.

With the exception of Acquire run with the search model, all of the current models that measure or predict interpretability (NIIRS) or performance (discrimination probabilities) assume unlimited viewing time and target fixation. They thus represent upper bounds on performance. A NIIRS 5 image theoretically allows one to "identify, by type, deployed SSM systems." This does not mean, however, that the analyst will necessarily find the deployed SSM, particularly if search time is limited. Similarly, a TTPF probability of recognition estimate of, say, 0.9, is valid only if the operator fixates the target within the allowable viewing time.

References

[1]   U.S. Government, *General Image Quality Equation (GIQE), Users Guide, Version 3.0*, Washington, DC: U.S. Government Printing Office, 1994.

[2] Ratches, J., *NVL Static Performance Model for Thermal Viewing Systems*, Report ECOM 7043, Ft. Monmouth, NJ: USA Electronics Command, April 1975.

[3] Eastman Kodak Company, *Physique (EOI) Math and Theory Document*, 1987.

[4] Eastman Kodak Company, *Physique (Radar) Formulation of the Model*, 1994.

[5] Mitchel, R. H., *SAR Image Quality Analysis Model*, Report 671034.3.X, Ann Arbor, MI: Environmental Research Institute of Michigan, 1974.

[6] Charmin, W. N., and A. Olin, "Tutorial-Image Quality Criteria for Aerial Camera Systems," *Photo. Sci. Eng.*, Vol. 9, No. 6, 1965, pp. 385–397.

[7] ARPA, *HAE UAV Concept of Operations*, Version 1.0, 1994.

[8] Imagery Resolution and Reporting Standards Committee, *General Image Quality Equation, Version 4.0, User's Guide*, 1996.

[9] Leachtenauer, J., et al., "The General Image Quality Equation," *Applied Optics*, November 1997, pp. 8322–8328.

[10] Berk, A., L. S. Bernstein, and D.C Robertson, *MODTRAN, A Moderate Resolution Model for LOWTRAN 7*, GL-TR-89-0122, Hanscom Air Force Base, MA: Air Force Geophysics Laboratory, 1989.

[11] Eismann, M. T., and S. D. Ingle, *Utility Analysis of High-Resolution Multispectral Imagery, Volume 3: Image Based Sensor Model (IBSM) User's Manual*, Ann Arbor, MI: Environmental Research Institute of Michigan, 1995.

[12] Fiete, R. D., and T. A. Tantalo, "Image Quality of Increased Along-Scan Sampling for Remote Sensing Systems," *Optical Engineering*, Vol. 38, No. 5, 1999, pp. 815–820.

[13] Fiete, R. D., "Image Quality and lFN/p for Remote Sensing Systems," *Optical Engineering*, Vol. 38, No. 7, 1999, pp. 1229–1240.

[14] Johnson, J., "Analysis of Image Forming Systems," in *Proc. Image Intensifier Symp.*, Ft. Belvoir, Virginia, October 1958, Warfare Vision Branch, Electrical Engineering Department, U.S. Army Engineering Branch and Development Laboratories, pp. 249–273.

[15] Rossell, F., "A Review of the Current Periodic Sensor Models," in *Proc. of IRIS Imaging*, 1981, p. 71.

[16] Lloyd, J., *Thermal Imaging Systems*, New York, NY: Plenum Press, 1975, p. 183.

[17] Sendall, R., and F. Rosell, "EO Sensor Performance Analysis and Synthesis TV/IR Comparison Study," Final Report AFAL-TR-72-374, Wright-Patterson Air Force Base, OH: Air Force Avionics Laboratory, April 1973.

[18] *FLIR'92 Thermal Imaging Systems Performance Model*, Ft. Belvoir, VA: U.S. Army Night Vision and Electronic Sensors Directorate, January 1993.

[19] Vollmerhausen, R., and R. Driggers, "NVTHERM: Next Generation Night Vision Thermal Model," *IRIS Passive Sensors*, 1999.

[20] D'Agostino, J., "Three Dimensional Analysis Framework and Measurement Methodology for Imaging System Noise," in *Proc. IRIS Passive Sensors Symp.*, March 1991.

[21] Webb, C., P. Bell, and G. Mayott, "Laboratory Procedure for the Characterization of 3-D Noise in Thermal Imaging Systems," in *Proc. IRIS Passive Sensors Symp.*, March 1991.

[22] Vollmerhausen, R., "Incorporating Display Limitations into Night Vision Performance Models," in *Proc. 1995 IRIS Passive Sensors*, Vol. 2, 1995, pp. 11–31.

[23] Vollmerhausen, R., "Minimum Resolvable Contrast Model for Image Intensified Charge Coupled Device Cameras," NV-96-Report 09, Ft. Belvoir, VA: U.S. Army CECOM Night Vision and Electronic Sensors Directorate, 1996

[24] Shumaker, D., J. Wood, and C. Thacker, *FLIR Performance Handbook*, Alexandria, VA: DCS Corporation, 1988.

[25] Howe, J., "Electro-Optical Imaging System Performance Prediction," in *The Infrared and Electro-Optics System Handbook, Volume 4*, Bellingham, WA: SPIE Press, 1993, pp. 91–116.

[26] D'Agostino, J. D., W. Lawson, and D. Wilson, "Concepts for Search and Detection Model Improvements," *SPIE Aerosense Conference*, April 23, 1997.

[27] Rosell, F. A., and R. H. Willson, "Recent Psychophysical Experiments and the Display Signal-to-Noise Ratio Concept," in L. M. Biberman (ed.), *Perception of Displayed Information*, New York, NY: Plenum Press, 1973.

[28] Lawson, W. R., T. W. Cassidy, and J. A. Ratches, "A Search Prediction Model," *SPIE Aerosense Conference*, April 23, 1997.

[29] Vollmerhausen, R., R. G. Driggers, and B. O'Kane, "Influence of Sampling on Target Recognition and Identification," *Optical Engineering*, Vol. 38, No. 5, 1999, pp. 763–772.

[30] Driggers, R., et al., "Laboratory Measurement of Sampled Imaging System Performance," *Optical Engineering*, Vol. 38, No. 5, 1999, pp. 852–861.

[31] Vollmerhausen, R. H., and R. Driggers, *Analysis of Sampled Imaging Systems*, Bellingham, WA: SPIE Press, 2000.

[32] Eastman Kodak Company, *Spectral Radiometric Measurement and Analysis Program*, AWS/TN-79-001, 1979.

[33] Oliver, C., and S. Quegan, *Understanding Synthetic Aperture Radar Images*, Norwood, MA: Artech House, 1998.

[34] Bhanu, B., and G. Jones, "Recognizing Target Variants and Articulations in Synthetic Aperture Radar Images," *Optical Engineering*, Vol. 39, No. 3, 2000, pp. 712–723.

[35] Nill, N. B., and B. H. Bouzas, "Objective Image Quality Measure Derived From Digital Image Power Spectra," *Optical Engineering*, Vol. 31, No. 4, 1992, pp. 813–825.

[36] Nill, N. B., and B. H. Bouzas, *Image Quality Assessment via Digital Power Spectrum Analysis*, MTR10921, Bedford, MA: Mitre, 1990.

[37]   Lubin, J., *A Methodology for Imaging System Design and Evaluation*, Princeton, NJ: David Sarnoff Research Center, 1995.

[38]   Lubin, J., "The Use of Psychophysical Data and Models in the Analysis of Display System Performance," in A. B. Watson (ed.), *Visual Factors in Electronic Image Communication*, Cambridge, MA: MIT Press, 1993.

[39]   Carlson, C., and R. Cohen, "A Simple Psychophysical Model for Predicting the Visibility of Displayed Information," *Proceedings of the Society for Information Display*, Vol. 21, 1980, pp. 229–245.

[40]   Barten, P. G. J., "The SQRI Method: A New Method for the Evaluation of Visible Resolution on a Display," *Proceedings of the Society for Information Display*, Vol. 30, 1987, pp. 239–243.

[41]   Ahumada, A., and A. B. Watson, "Equivalent Noise Model for Contrast Detection and Discrimination," *J. Opt. Soc. Am.[A]*, Vol. 2, 1985, pp. 1133–1139.

[42]   Snyder, H. L., et al., "Digital Image Quality and Interpretability: Data Base and Hardcopy Studies, *Optical Engineering*, Vol. 21, 1982, pp. 14–22.

[43]   Rohaly, A. M., et al. (eds.), *Final Report from the Video Quality Experts Group on the Validation of Objective Models of Video Quality Assessment*, December 1999.

[44]   Martens, J., and L. Meesters, "The Role of Image Dissimilarity in Image Quality Models," *Proceedings of the IS&T/SPIE Conference on Human Vision and Electronic Imaging IV*, Vol. 3644, San Jose, California, SPIE, 1999, pp. 258–269.

[45]   Avadhanam, N., and R. Algazi, "Evaluation of a Human Vision System Based Image Fidelity Metric for Image Compression," *SPIE Conference on Applications of Digital Image Processing XXII*, Denver, Colorado, 1999, pp. 569–579.

[46]   Berfanger, D. M., and N. George, "All-digital Ring-wedge Detector Applied to Image Quality Assessment," *Applied Optics Information Processing*, Vol. 39, No. 23, 2000, pp. 4080–4097.

[47]   Fairhurst, A. M., and A. H. Lettington, "Method of Predicting the Probability of Human Observers Recognizing Targets in Simulated Thermal Images," *Optical Engineering*, Vol. 37, No. 3, 1998, pp. 744–751.

[48]   Imagery Analysis Division, *Multispectral IIRS Development*, Washington, DC: National Exploitation Laboratory, 1995.

[49]   Martin, L., J. Vrabel, and J. Leachtenauer, "Metrics for Assessment of Hyperspectral Image Quality and Utility," *ISSSR Proceedings*, Las Vegas, Nevada, 1999. CD-ROM, November 2000.

[50]   Martin, L., et al., "Image Quality Evaluation of AOTF and Grating Based Hyperspectral Sensors," *ISSSR Proceedings*, Las Vegas, Nevada, 1999.

11

Sensor Performance Conversions

Target acquisition sensor (TAS) performance [1–5] is currently modeled using the probability of discrimination technique. This method is described in Section 10.1.2. S&R sensor performance is frequently modeled using the NIIRS and the GIQE [6,7] as described in Sections 5.4.1 and 10.1.1. TAS sensors are fielded on a variety of weapon platforms including tanks, personnel carriers, fire-control systems, rotary- and fixed-wing aircraft, and low-flying UAVs. S&R sensors are typically fielded on high-altitude aircraft, high-flying UAVs, and commercial satellites. Figure 11.1 shows the altitude (around 12,000 ft) where sensors transition from tactical to S&R, as well as the methods of design and analysis change.

There are a number of major efforts under way that will make S&R sensor data available for target acquisition purposes (and vice versa). The National Imagery and Mapping Agency (NIMA) is one of these efforts. NIMA's mission is to ensure the timely dissemination of digital imagery to forces in the field. The military services are moving to a defense network in which imagery from any platform, including commercial, is available to an S&R or TAS user. In the near future, both government and commercial sensor engineers and users will have to understand each other's nomenclature and sensor descriptions. A difficulty to overcome is that the TAS and S&R communities use different imaging system models and, in general, do not share information. Figure 11.1 shows a number of military imaging platforms and the sensor description technique (NIIRS or probability of recognition (Pr)) employed for each sensor.

The primary TAS performance measure is the probability of visual target discrimination, which involves detection, recognition, and identification.

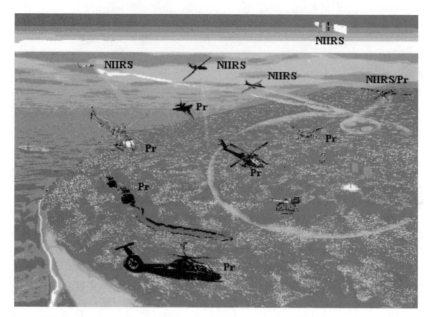

Figure 11.1 Platforms and sensor performance descriptions.

The discrimination task is modeled using the target-to-background contrast, target dimensions, atmospheric characteristics, and sensor performance. The TAS technique is similar to the techniques used in linear circuit theory and is described as a "systems" approach to sensor performance modeling. The model is extremely flexible in that components such as optics, scanners, detectors, etc., can be added or subtracted from the system and the performance can be quickly determined. The overall performance is described using Johnson's criteria and the sensor's MRT parameter. Johnson's criteria describe the number of bars (in a bar pattern) that must represent an object for a particular level of discrimination. The MRT parameter describes the sensor's sensitivity as a function of bar pattern spatial frequency.

S&R systems are described and tasked in terms of NIIRS performance. The NIIRS level defines the level of exploitation that can be achieved for each of several classes and sizes of objects. The GIQE is the sensor model that predicts NIIRS performance as a function of design and operating parameters (e.g., scale, resolution, and SNR). Given a predicted NIIRS level, a user can determine whether the sensor is useful for a particular collection task. Similarly, given a set of NIIRS requirements, a designer can determine the system design characteristics needed to satisfy those requirements. The GIQE is an empirical model that was developed using statistical analysis of imagery analyst responses.

Sensors, such as those on Predator and Global Hawk, will soon be used for both S&R and target acquisition purposes. A conversion was recently developed for the NIIRS and Pr that depends on sensor resolution and sensitivity. Such a conversion allows the user and sensor designers to utilize and design precious sensor assets correctly. For example, a tactical user would not be required to guess whether a sensor with NIIRS 4–rated imagery would be able to identify a tank at a given range with a high probability.

A performance conversion was first developed for resolution-limited sensors. Last year, however, a performance conversion was developed that included both resolution and sensitivity.

In this chapter, we begin by showing a relationship between NIIRS and the Johnson criteria. We then describe a method for converting NIIRS and Pd, Pr, and Pid using both the TAS models and the S&R models. Next, a straightforward calculation that does not require sophisticated computer code is described. Finally, results are given where both target acquisition specialists and imagery analysts were shown the same imagery to determine an empirical relationship.

11.1 Johnson Criteria and NIIRS

The Johnson criteria are used as input to the TAS models. These criteria define the level of detail necessary to achieve a desired level of discrimination performance. The NIIRS defines the level of interpretability of an image, sometimes in terms similar to the Johnson criteria. For some of the NIIRS criteria, it is possible to define the minimum size of the referenced objects from published sources. Additionally, for a given system, it is possible to define the relationship between NIIRS and GSD using the GIQE. This provides a rough basis of comparison between the Johnson criteria and NIIRS.

For purposes of comparison, 11 visible NIIRS criteria over the range of NIIRS 3 to 6 were selected. The criteria were selected because the minimum dimension (in the ground plane) of the referenced objects could be easily defined. Objects included rail cars, vehicles, ships, and aircraft. The minimum dimension of the object was divided by the Johnson criterion appropriate for the level of discrimination. For example, if the NIIRS criterion called for recognition, the dimension of the object was divided by 3.0 (the Johnson criterion for recognition). This dimension was compared to the GSD value corresponding to the NIIRS level.

Results are shown in Figure 11.2. The dashed line is the regression fit to the data; the solid line represents a 1:1 relationship. The log GSD term accounts for 93% of the variance in cycle width. Although this is a strong

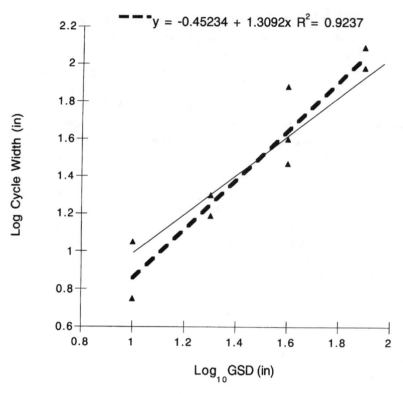

Figure 11.2 Relationship between Johnson criteria and NIIRS in GSD space.

relationship, there is considerable variation for individual criteria. It was noted that objects defining a given NIIRS criterion could show substantial dimensional variability (up to 2:1). As discussed in Chapter 7, there is not always a well-defined relationship between overall target size and the size of the cue needed for recognition. Finally, note that the data in Figure 11.2 equates cycle width and GSD on almost a 1:1 basis as opposed to the expected 2:1 relationship. There are two reasons for this. First, the Johnson criteria define performance at the 50% level; the NIIRS is more likely a higher (but undefined) level of performance. Second, both the NIIRS and Johnson criteria are related to some underlying level of image quality that has never been equated. This deficiency indicates the need for more rigorous approaches in relating S&R and TAS modeling.

11.2 Conversions with Models

The relationship between NIIRS and the probabilities of discrimination (detection, recognition, and identification) should be a constant one because

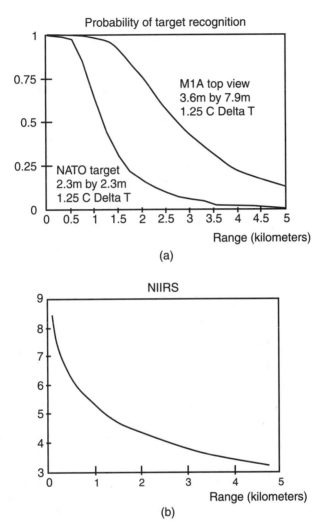

Figure 11.3 (a) TAS and (b) S&R performance curves for the midwave example.

both are an image exploitation measure. The TAS and S&R models, however, are not sufficiently mature; they still require a good amount of development as the relationship may vary slightly from sensor to sensor. A straightforward method for developing an NIIRS to Pd, Pr, Pid conversion is to run the FLIR92/NVTherm/ACQUIRE model on a sensor as applied to a particular target and to run the GIQE model on the sensor and compare the results. Figure 11.3 shows the performance curves for the midwave sensor given in Section 10.1.2. The GIQE was applied to the sensor, and that curve is shown in (b). First, note that the Pr changes with different target size (a). The NIIRS is independent of target size because the target characteristics are inherent

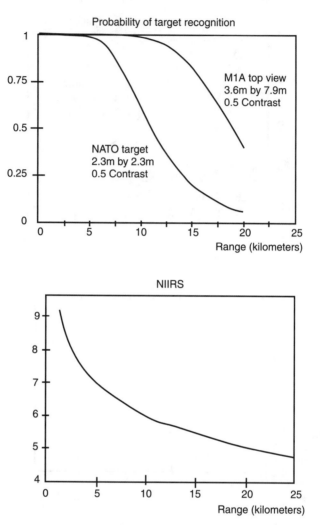

Figure 11.4 TAS and S&R performance curves for the EO example.

to the definition of the NIIRS. These two performance curves are both plotted as a function of range. Therefore, they can be compared at each range to relate the two performance parameters. Figure 11.4 shows the two performance conversions for the EO example presented in Section 10.1.2. The curves in the top part were generated with an MRC model that was applied to a contrast target with the acquisition model. The curve in the bottom part was obtained by applying the EO GIQE equation to the sensor.

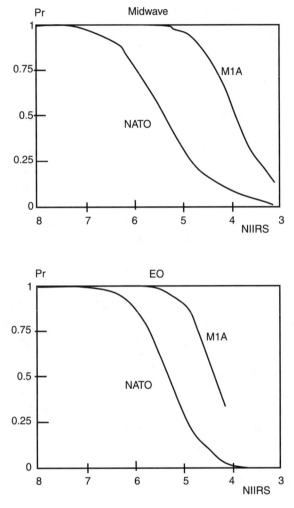

Figure 11.5 NIIRS and Pr performance conversions for the midwave and EO sensors.

Using range as a common denominator, these examples can be mapped to relate the probability of recognition to NIIRS. Figure 11.5 shows these mappings.

From Figure 11.5, an NIIRS 5.5 gives a Pr of around 0.5 for a standard NATO tank. An NIIRS 4 gives a 0.5 Pr for the top view of an M1A tank. Because the GIQE model is a little different for EO systems, just under an NIIRS 6 gives a 0.5 Pr for a standard NATO target and an NIIRS 5 gives a 0.5 Pr for the top view of an M1A tank.

11.3 Comparing Sensor Performance Conversions

In the previous section, we showed that a conversion can be obtained by applying the TAS and S&R models to a sensor and comparing the results. In this section, we compare three conversions that provide a probability of identification as a function of the NIIRS level. The first conversion was obtained by applying both models (FLIR92 for the TAS values and the GIQE for the S&R values) on the same sensor and plotting the results as in the previous section. The second is a closed form mathematical expression for probability of identification as a function of NIIRS level. Finally, an empirical expression is presented. All of these conversions assume the same sensor. The sensor that is modeled here for comparative purposes is a midwave (3–5 μm) thermal imager with an f-number of 2 and a focal length of 9.175 cm. The detector was 20 × 20 μm with a sampling rate of two samples per detector dwell. There were six levels of display reconstruction blur to vary target identification probabilities. The display blurs were Gaussian and, in sensor space, were 0.197, 0.295, 0.442, 0.590, 0.885, and 1.19 mrad. The image on the monitor subtended an angle to the observer of 18.5 degrees. The sensor field of view was 3.2 degrees for a magnification of 5.8. The detectivity was set to 1×10^{11} Jones, and the integration time was 13.33 ms for a noise equivalent temperature difference of 0.057 K.

The sensor described above was applied to a set of confusable targets. The targets were all tracked vehicles as shown in Figure 11.6. These targets were chosen from a database of targets in order to force the discrimination level to identification. The contrast of the targets was taken to be 2°C. A representative dimension (assuming a square target) was taken as the average of the geometric mean of target height, width, and length. This representative target dimension was 3.88m. Acquisition calculations were performed using this critical dimension and estimated contrast.

The tactical acquisition model was applied to the sensor described above. The GIQE was also applied to this sensor in order to determine the NIIRS estimates of the sensor output imagery. Both models were applied to the sensor over the range of display blur described in the previous sensor description, and a probability of identification was estimated for a target positioned at 850m from the sensor. This range was used because it corresponded to the images used in the empirical measurements of the probability of identification by participating U.S. soldiers and the NIIRS estimates of NIMA's imagery analysts. The results of the probability of identification were determined as a function of the NIIRS estimates.

Because Johnson's criteria use a limiting frequency concept, where the limiting frequency is the result of the highest frequency resolvable at the

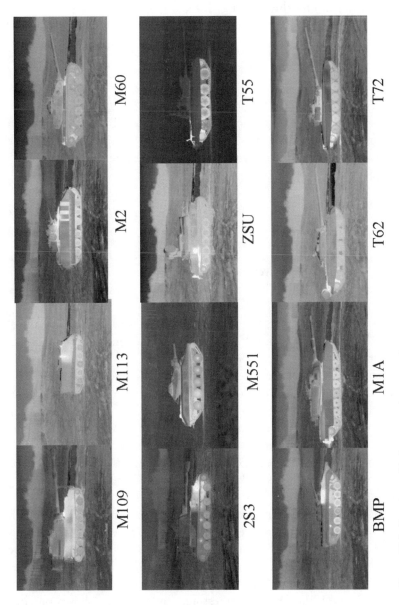

Figure 11.6 Targets used in the conversion comparison.

apparent target differential temperature, we applied a similar concept to the GSD. That is, a limiting GSD is determined, where the smallest resolvable GSD is a function of both resolution and sensitivity. Using the GIQE as the guiding relationship for this limiting GSD,

$$GSD_{min} = 10^{\left[\frac{10.751 + b_{\log 10}RER - NIIRS - 0.656H - 0.344G/SNR}{a}\right]} \qquad (11.1)$$

where a and b are as in (10.2).

This GSD corresponds to the equivalent GSD seen by an observer when viewing an image with both blur and noise. Using this limiting GSD, the number of cycles across the target dimension can be determined:

$$N = \frac{d_c}{2GSD} \qquad (11.2)$$

This is the same number of cycles seen across the target using the MRT. This value is input to the TTPF in order to determine the probability of target detection, recognition, or identification.

Ten U.S. Army soldiers were trained to recognize all of the targets described in Figure 11.5. Personnel were trained on the multimedia target recognition training program developed by NVESD, ROC-V, which is currently used to train soldiers and Marines on IR target recognition. The imagery in the training program was pristine, high-fidelity imagery. Each individual trained at their duty station or on-site at the Night Vision Perception Facility until they were able to attain a recognition criterion of 95% correct on a post-test composed of the 12 targets at three orientations and two azimuths (ground-to-ground and 15-degree elevated). Reaching this criterion depended upon the particular observer's background and natural ability; some requiring a few days of training and others only a few hours.

The soldiers were then required to identify tanks that were seen through the sensor process described above. All of the targets were processed through simulated sensors. Each probability cell included 48 targets of random type and orientation. The experiment was conducted at McDill Air Force Base, Tampa, Florida, and at the Night Vision Laboratory, Ft. Belvoir, Virginia, in May and June 1998. During the testing, observers were shown one image at a time, in balanced blocks, with random presentations. The responses were timed, but the observers were given unlimited time to make their recognition choices from a menu consisting of the 12 target choices in a forced choice manner (the observer had to make a choice on each trial). The test required approximately 2–3 hours of an observer's time.

The imagery was sent to NIMA for NIIRS analysis. Of the six sensor configurations, only four of these configurations were considered usable for NIIRS analysis. The larger display blur cells did not leave enough information for ISR analysis, and the field of view was not large enough for lower NIIRS estimations.

The graph of the three comparisons is shown in Figure 11.7. It is clear that the performance runs of the two imaging system models match the results of the conversion equation. Both of these methods appear to match the empirical comparison in a reasonable manner. It is not apparent from this graph that the comparison holds over a range of sensitivity values. Therefore, the noise in terms of the sensor NETD was adjusted using the detectivity of the sensor. The noise was imposed on the sensor configuration associated with the 0.59-mrad reconstruction blur so that the noise could vary both the Pid and the NIIRS around a mid-level resolution point. The NETD was adjusted to be 0.01, 0.1, and 1 K to give an SNR of 174, 17.4 and 1.74 at a range of 850m with a broadband Beer's Law transmission approximation of 0.85 per kilometer. The Pids associated with these SNRs were 60%, 44%, and 24%, and the NIIRS were 5.5, 5.4, and 5.2. These values follow the same trend shown in Figure 11.4, so the sensitivity dependency appears to hold within the relationship given by both the models and by the resolution-sensitivity equation given in Section 11.2.

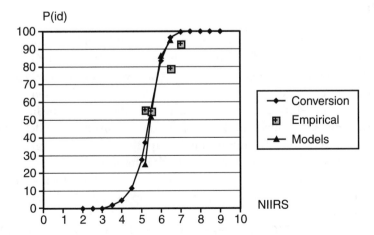

Figure 11.7 Conversion results.

11.4 Performance Conversion as a Function of Target Size

Because the probabilities of discrimination depend heavily on target size, but the NIIRS has target size inherent to the definition, the relationship between Pd, Pr, and Pid and NIIRS is a function of target size. Consider the relationships shown in Figure 11.8. These relationships were run 3 to 4 years ago and since then the GIQE has been modified and the relationships shown have matured. A lower NIIRS level provides the probabilities shown. The figure, however, illustrates the conversion as a function of target size.

Figure 11.8 shows the NIIRS and probability of detection, recognition, and identification values for four targets: the standard NATO target, a top view of an M1 Abrams, a top view of an F-15 aircraft, and a top view of a B-52 aircraft. Note here that aircraft recognition was assumed to follow the cycle criteria described previously (the cycle criteria provided was developed with ground targets and side views). Also, as the target becomes larger, a much smaller NIIRS is required to perform detection, recognition, and identification tasks. For example, consider the NATO target graph and compare it to the M1 Abrams top view graph. They are both tanks, but the NATO target gives an NIIRS 7.1 estimate for a 50% probability of recognition. The M1 Abrams would be recognized with the same probability with just under an NIIRS 6. While the NATO target represents the front view of a small tank, the top view of an M1 Abrams is significantly larger than the NATO target. This difference in target size accounts for the difference in the required NIIRS level for the same recognition probability.

11.5 Summary

There are a number of differences between the GIQE and probability of discrimination techniques. First, the GIQE currently defines the GSD in the ground plane and is limited to look angles (angle between ground plane and line of sight to the sensor) of 15 degrees or greater. There is evidence to suggest that GSD should be defined in the plane orthogonal to the line of sight at lower angles, but sufficient data is currently unavailable to model the transition. The probability of discrimination approach uses the target's critical dimension, which accounts for the transition from top to side views. These effects are not shown here because the target size was held constant (as in the side view of the NATO target and the top view of the M1A target) to achieve a comparison.

Second, the GIQE currently predicts NIIRS for hardcopy (film) imagery. Performance on a good quality softcopy (CRT) display provides an

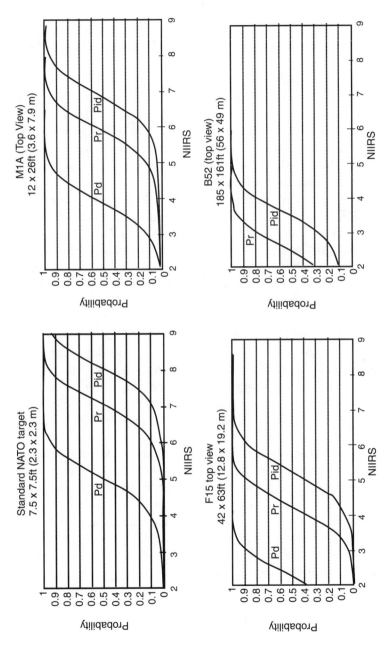

Figure 11.8 Performance conversions as a function of target size.

increase of around 0.2 NIIRS better than predicted. A poorly designed or maintained CRT, however, can result in lower than predicted interpretability of 1 NIIRS or more. The probability of discrimination is typically performed on softcopy display and there is no explicit definition of the display. It is important to note, therefore, that the prediction is valid only for the display used for prediction.

Some general trends can be stated for both the Pr and the GIQE (NIIRS) approaches to sensor performance. The ability to recognize an object increases with increasing target size for a given acquisition system and set of operating conditions. For the NIIRS/GIQE characteristics, the target size relationship is implicit in the criteria definitions but is not directly evident from the GIQE predictions. Therefore, the relationship between Pr and NIIRS changes dramatically with target size. Since the IR and EO system examples of Pr and NIIRS were both plotted with range, we can relate the Pr to the NIIRS estimates for each sensor. Figure 11.7 shows both the IR example relationship and the EO example relationship. Note that using the top view of the M1A tank, a 50% Pr for the IR system occurs at an NIIRS of around 5.5 for the NATO target and around 4 for the M1A target. For the EO system, the 50% Pr occurs at around NIIRS 6 for the NATO target and just over an NIIRS 5 for the M1A target. Recall that the GIQE is different for the IR and EO systems.

The NIIRS tables are also different for IR and EO systems. Another difference is the TAS definitions of recognition were developed for the discrimination of smaller tanks, such as a T-62, from trucks and nondiscrimination of large bridge-crossing equipment from larger and even smaller tracked vehicles. Next, Johnson's criteria were developed for low aspect (look) angles, whereas the NIIRS was developed with large aspect angles. It is possible that the larger aspect angles require a different amount of information for discrimination tasks. There is evidence from preliminary experiments at NIMA that three cycles across the target dimension give a higher than 50% Pr. This would account for some of the recognition capacity at lower NIIRS levels. Finally, Johnson's original criteria were given as discrimination definitions for detection, orientation, recognition, and identification. The NIIRS tasks are defined in terms of detect, distinguish, differentiate, and identify. The levels of discrimination are not identical, so there are some discrepancies when comparing task-oriented performance.

References

[1]   Johnson, J., "Analysis of Image-Forming Systems," in *Proc. Image Intensifier Symposium*, Ft. Belvoir, VA: Warfare Vision Branch, Electrical Engineering Department, U.S. Army Engineering Development Laboratories, October 1958, pp. 249–273.

[2] Rossell, F., "A Review of the Current Periodic Sensor Models," in *Proc. of IRIS Imaging*, 1981, p. 71.

[3] Lloyd, J., *Thermal Imaging Systems,* New York, NY: Plenum Press, 1975, p. 183.

[4] Sendall, R., and F. Rossell, *EO Sensor Performance Analysis and Synthesis (TV/IR Comparison Study),* Final Report, AFAL-TR-72-374, Wright-Patterson Air Force Base, OH: U. S. Air Force Avionics Laboratory, 1973.

[5] Ratches, J., *NVL Static Performance Model for Thermal Viewing Systems,* Ft. Monmouth, NJ: USA Electronics Command Report ECOM 7043, 1975.

[6] Leachtenauer, J. C., "National Imagery Interpretability Rating Scales: Overview and Product Description," in *APRS/ASCM Annual Convention and Exhibition Technical Papers: Remote Sensing and Photogrammetry,* Baltimore, MD: American Society for Photogrammetry and Remote Sensing and American Congress on Surveying and Mapping, 1996, Vol. 1, pp. 262–272.

[7] Leachtenauer, J., et al., "General Image Quality Equations: GIQE," *Applied Optics*, Vol. 36, No. 32, November 1997, pp. 8322–8328.

12

Conclusions and Future Directions

Despite over 40 years of research on the prediction of S&R system performance, many issues remain. In this chapter, we review some of the issues and suggest potential avenues of further research.

The Night Vision and Electronic Sensors Directorate (NVESD) probability of discrimination and general image quality equation (GIQE) parameter-based models predict performance for TAS and S&R EO and IR systems with good accuracy, provided that certain conditions are met. The models predict for a sample of imagery or targets on average, not individual targets or images. Sampling and bandwidth compression effects are not explicitly treated, and the models predict only for single-channel monochrome imagery. The NVESD model defines display luminance and geometry; the GIQE assumes an optimized hardcopy display. Although Physique in theory overcomes many of these limitations, validation data has not been published and thus its utility remains to be demonstrated.

The parameter-based models treat spatial effects (scale, resolution, contrast, and noise) but do not deal with the spectral and temporal dimensions of imaging systems. Further, they deal only with EO and IR imagery and assume that the observer fixates the target, thus generally ignoring the process of search.[1] SAR is largely ignored in the context of predicting human observer performance, at least for high-resolution systems.

The image-based models such as image quality model (IQM) and the Sarnoff just-noticeable difference (JND) model have been demonstrated only

---

1. The exception is the search term developed for the NVESD Acquire mode. Significant prediction errors have been shown, however.

with visible data for use in National Image Interpretability Rating Scale (NIIRS) prediction. Despite its lack of demonstrated success in predicting video double stimulus continuous quality scale (DSCQS) ratings, the JND or similar models appear to offer a better long-term approach than peak signal-to-noise ratio (PSNR). None of the existing models has been shown to deal effectively with search. Finally, little work has been done in the spectral and temporal domain.

12.1 Spectral Domain

Multispectral and hyperspectral systems provide, or potentially provide, spectral signature data that can be used as a recognition cue. In theory, single-pixel identification is possible based on such signatures. For a variety of reasons, however, this is seldom true in actual practice. Because of sampling effects, single pixels are seldom homogeneous. Typically, a minimum of nine pixels is required to unambiguously achieve a single "pure" pixel. A second issue is the variability of spectral signatures. Items of the same apparent color (green vegetation, for example) can have different spectral signatures. Signatures of the same object (e.g., a tank) can vary because of differences in aging of paint, dirt, and lighting. Just as some objects have unique appearances and others do not, some objects have unique spectral signatures and others do not. Finally, the human visual system does not do well at color matching [1]. Size, luminance, surround luminance, and hue and saturation of the surround can all affect the appearance of colors. For this reason, signatures must be unique to be visually useful. In some cases, improved uniqueness can be achieved through processing, but only at the expense of reducing color contrast for some other signatures.

Increasing the spectral bandwidth of a system offers additional chances for a unique signature. The addition of an SWIR band was shown to increase MS IIRS ratings by 0.4 scale units [2]. The increase is offset, however, by the difficulty of operating in this wavelength band. There is typically a 2:1 (or more) resolution penalty in moving from the visible and NIR to the SWIR. An even greater penalty (4:1) is incurred in moving to the MWIR. The value of additional bands in a defined spectral range is less clear-cut. As bands are narrowed, S/N generally decreases, or conversely, S/N is maintained at the expense of resolution (GSD). The benefit of increasing the number of bands is in part a function of the spectral characteristics of the objects to be discriminated. When properly placed, a relatively small number of bands can provide discrimination. In general, evidence suggests that there is greater advantage to moving into the SWIR as opposed to increasing the number of visible bands. Hyperspectral imaging systems offer the ultimate in spectral signature sensitivity, but again at the expense of resolution.

At the present time, it appears that the GIQE, with some modification, including calibration to the MS IIRS, can satisfactorily predict the performance of a four-band (visible and NIR) MSI system. The NVESD/PD model, with appropriate treatment of the MRC model, can also be used. The GIQE could also predict for the additive effect of an SWIR band. Sufficient data, however, does not appear to be available for the NVESD/PD model to predict an additive benefit of an SWIR capability, if in fact it even exists. In both cases, only spatial attributes are or would be treated. The JND model is theoretically capable of dealing with spectral quality degradation, although calibration to the MS IIRS would be required. Any system approaching a hyperspectral capability will require a new approach to deal with spectral and radiometric issues.

The multispectral case can be extended beyond the visible and SWIR to the MWIR and LWIR and even to radar. No models currently exist to predict the performance of such systems, and it is not clear that such systems would be particularly useful in the TAS and S&R application. The loss in resolution incurred at the longer IR wavelengths may not offset the increase in spectral diversity. Adding a radar capability vastly increases system complexity, and again, the increase in information may not warrant the added complexity. In both cases, answers are not yet available.

Finally, radar offers several possibilities in terms of multi-image solutions. These include multifrequency, multiview, and multipolarization imagery. The NASA/JPL AirSAR system [3] collects C-, L-, and P-band data at four polarizations (VV, VH, HH, HV). These types of data can be very useful in analysis of clutter from vegetation, and in the case of polarization, they can provide different returns depending on the orientation of the reflectors. As the number of images increases, however, the time required for collection and analysis increases and for direct viewing quickly becomes infeasible. Thus, although multifrequency and multiple polarization systems offer theoretical benefits, their practical benefits can not be determined in the absence of specific processing/display algorithms.

Multiview imagery consists of imaging an area from multiple aspect or orientation angles. Again, provided that the imagery can be displayed in a meaningful way, some benefit may occur. The benefit, however, must be traded against the cost of acquiring multiple images (as opposed to acquiring a single higher resolution image). In the Spotlight mode of imaging, resolution is improved by keeping the target in the radar beam for a longer time

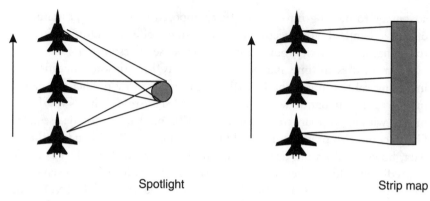

Spotlight   Strip map

Figure 12.1 SAR imaging modes.

than would occur in strip-mapping (Figure 12.1). An alternative approach is to noncoherently combine multiple (lower resolution) images of the target.

Models do not exist to predict information extraction for multi-image SAR systems. Models do exist, however, to predict the properties and characteristics of such imagery [3]. Again, the value of such techniques remains to be demonstrated in the S&R/TAS domains.

12.2 Temporal Effects

Although temporal domain effects can be quite important in S&R applications, they have not been modeled in the sense of information extraction. Surveillance in particular entails analysis of change over time. The change may be some type of human activity, or it may be natural activity such as crop growth. A military base may be repeatedly imaged to determine indications of future (or recent) actions. A manufacturing plant may be repeatedly imaged to estimate production rates.

In most, if not all, cases there are two elements to the change. The first is the item of interest. In the example of crop growth, it is some type of plant. In the case of the manufacturing plant, it is raw materials and the results of the production process, either directly or indirectly. For example, it might be possible to count vehicles at a tank production facility, but only box cars or trucks at an electronics factory. The second element is the time scale of the activity. In many cases, this is known a priori or can be determined after some period of observation. Growth occurs over a period of weeks and months, as does most construction activity. Aircraft departures occur over periods of minutes, although preparation for departure may take several minutes to hours. Loading and unloading of ships takes several hours. Production rates vary in duration as a function of the unit of production and the size of the plant. In every case, the required rate or frequency of surveillance can be established, perhaps after some trial and error. The modeling process then simply requires that the item(s) of interest be detected or recognized to the necessary level of detail.

A second type of temporal effect occurs when the motion of a target object provides information. The movement of a vehicle over time provides data on speed and acceleration if the sampling rate is sufficient. The rotation of an antenna provides multiple looks at the antenna over time. High-speed video may provide weapon effects information in terms of both the behavior and effect of the weapon. As in the previous discussion of change detection, there are at least two elements to the problem—the spatial and the temporal. The temporal aspect is generally predictable; the spatial aspect is predictable for some problems but not others (e.g., weapons effects). Here, a trial and error approach may be required. A third element, which may or may not come into play, is the element of angle diversity. If the angle diversity is deterministic (e.g., a rotating or nodding antenna), it can be predicted and information obtained from the temporal domain used to advantage. If it is not predictable (e.g., a vehicle turning through a sequence of images), use of the temporal domain is likely to offer only random advantage.

In the case of the temporal domain, we conclude that additional information on the temporal behavior of activities may be useful, but there is probably not a need for models in the sense presented in this book. Use of the spatial domain models with sufficient sampling frequency should result in a high probability of information satisfaction.

12.3 Search

Despite the numerous studies performed on search, there do not appear to be any comprehensive successful models of the process. The VISDET model [4] discussed in Chapter 9, although reasonably comprehensive, was shown to overpredict performance [5]. The VISDET model assumed a random search pattern, when, in fact, search has been shown to not be random. A target contrast, a target/clutter similarity, and an observer vigilance term were included in the model. The Acquire search model assumes a constant fixation time and bases the probability of detection on static detection probability. In the context in which it is used (TAS), the model may be adequate, but it does not appear adequate for the S&R scenario. The assumption of random search and a constant average fixation time in particular are limiting.

In addition to the influence of objects in peripheral vision, scene or target knowledge, as well as apparent search biases, shown over 40 years ago by Enoch and Fry [6], contribute to the nonrandom nature of search. The Enoch and Fry data also showed that both image scale and quality impacted fixation time and saccade distance. They also showed the biasing effect of scale in an oblique image. Observers tended to avoid the regions of smaller scale, even though there was no relationship between target location and scale. Display size also affected fixation duration and saccade length. Because edges are important for recognition, it would seem that any obscuration of edges (vegetation, camouflage) would have an effect on search [7]. Finally, large differences in search performance exist among individuals and as a function of the instructions provided to persons performing search [4,8].

In its simplest form, the probability of detection in search is a function of the probability of fixating and recognizing a target. Again in a simple sense, the probability of fixating a target is a function of time available for search, average fixation time, and average saccade distance. Note that use of the term "average" implies variability about the mean. Unfortunately, in the S&R context (as opposed to, for example, open water search and rescue), fixation points are not uniformly distributed, and this contributes to the difficulty in predicting search performance. Depending on the target and the knowledge of the analyst, the image can be partitioned into areas differing in the likelihood of target occurrence. A greater concentration of fixations should logically be associated with areas of greater likelihood of target occurrence. Arguing against this logic is the analyst's tendency to concentrate fixations in the center of the display, to avoid smaller as opposed to larger scale areas of the display, and to be directed by attention-getting cues (e.g., bright objects) in the visual periphery. Fixation durations and saccades are affected by image quality, instructions, and observer differences [4].

If we can manage to predict the probability that the target will be fixated, we still must predict the joint probability that the target will be recognized as a target. Here we need to deal with issues of target conspicuity (size, shape, contrast, complete edges, lack of confusing objects) and issues related to signal detection theory. A host of factors can affect the analyst's decision criterion (β). Some of these factors affect individual analysts and some affect a group as a whole. For example, the perceived likelihood of a target being present affects β and hence the actual detection rate [9].

We thus conclude that considerably more work is required before accurate models of the search process can be developed. In particular, the use of real-world imagery and recording of eye fixations appear to be necessary. The VISDET and Acquire models appear to be a beginning for the TAS scenario, but further development is needed. In the case of the S&R scenario, no useful models appear to exist. Thus, both the GIQE and the static NVL models can be thought of as predicting some upper bound on performance. They both predict a level of image quality sufficient to perform the necessary task, provided that the target has been fixated and is not otherwise obscured (for example, with vegetation or camouflage).

12.4 SAR

Empirical or theoretical models that predict either NIIRS or probabilities of target recognition for current high-resolution SAR imagery do not currently exist. The RTQF was developed for coarser resolution data and was not fully validated.

The problem of predicting the performance of analysts viewing SAR imagery does not appear difficult if well-trained observers are available. The dependence of target signatures on imaging aspect angle [3] appears to present a challenging problem for the observer because of the need to learn and remember all of the signatures for a wide variety of target objects.

It is clear that a resolution term similar to the GSD term used in the GIQE is the first requirement. In the case of SAR, the -3-dB impulse response is the conventional resolution measure. The two-dimensional projection in the ground plane would appear to be the logical counterpart to the two-dimensional GSD. Again, following the logic of the GIQE, a logarithmic relationship with NIIRS would be expected.

The ability to separate target and background returns is a function of both clutter and noise. Again using the GIQE as an example, a term analogous to SNR is needed. The GIQE provides two alternatives for the treatment of SNR. In one, typical (and constant) target and background reflectance or radiance values are used. In the other, actual measured values are used. With SAR, the analogy to the use of a single target and background would be the use of a target with defined RCS and a particular background type. The σ° of a defined clutter varies as a function of range, becoming smaller at increased range. The RCS data from Table 6.4 is plotted as a function of range in Figure 12.2. The relationship is generally described by a second-order polynomial. From (6.18), SCR is independent of range for a given target RCS, but detection is also affected by noise. Noise can degrade with range, particularly for airborne systems [3]. The degradation is nonlinear. Consequently, the detectability of a target will vary with range depending on the level of noise and the CNR. As CNR decreases, the relative effect of noise will increase.

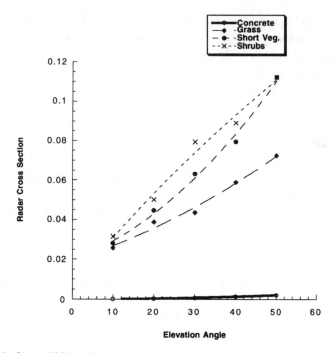

Figure 12.2 Clutter RCS (in dB) versus look angle.

For high-resolution SAR, targets are made up of several reflectors, and some level of recognition is required for detection. For example, returns from a vehicle must, at a minimum, be recognized as those from a vehicle as opposed to vegetation or some other manufactured object not of interest. Target returns vary as a function of imaging geometry and thus so must target cross-section. Effective target cross-section thus varies as a function of horizontal orientation and vertical aspect angle. Because horizontal orientation is generally not known a priori for many targets, orientation is perhaps best treated as a random variable and some average value of target RCS used at a given vertical aspect angle. Vertical aspect angle is a function of collection geometry and is known a priori.

For a given range and look angle, a system with a defined noise level, and a defined target and background type, an average target-to-background contrast exists that is equivalent to the SNR term used in the GIQE. This contrast, along with a measure of resolution, is the basis for predicting the probability of recognition. In the case of NIIRS prediction, the combination of system noise level, background type (clutter RCS), and system resolution defines a level of image quality relatable to NIIRS. In a conceptual sense,

$$\text{NIIRS} = k + a \log IPR_{Area} + bLA + cLA^2 + dRg + eRg^2 \quad (12.1)$$

where IPR_{Area} is the area of the −3-dB impulse response projected into either the ground plane or the orthogonal plane depending on the application, *LA* is look angle, and *Rg* is sensor-to-target range. Under some circumstances (flat earth, constant vehicle altitude), *LA* and *Rg* are redundant. Alternatively,

$$\text{NIIRS} = k + a \log IPR_{Area} + bTBR \quad (12.2)$$

where IPR_{Area} is as in (12.1), TBR is the ratio of target-to-background (clutter plus noise) RCS. TBR varies with range and elevation angle.

One additional measure of quality to be considered is the impact of sidelobes. When sidelobes are sufficiently high, they can provide returns that interfere with other target returns and thus potentially degrade recognition based on patterns of individual returns. Figure 12.3 shows an example. The strong sidelobes at "A" potentially interfere with other returns on the same vehicles. Most of the vehicles in the scene, however, do not exhibit noticeable sidelobes. For a well-designed system where sidelobes are controlled, sidelobes should thus have little effect on information extraction.

The GIQE additionally contains terms related to edge sharpness and noise gain as well as edge overshoot. Because SAR imagery generally consists of point returns, the edge overshoot term does not seem appropriate and the edge sharpness (RER) term would appear to be captured in the IPR term.

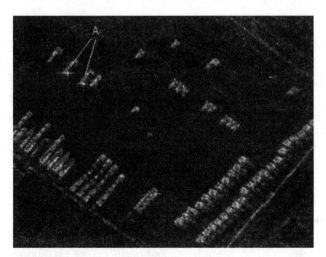

Figure 12.3 SAR sidelobes (courtesy of TESAR Program).

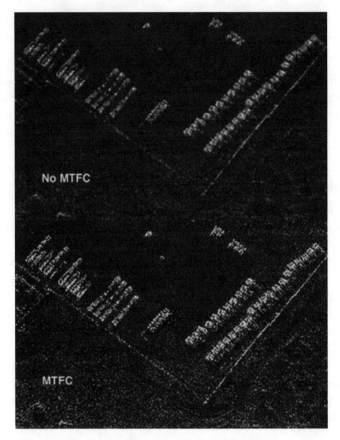

Figure 12.4 Effect of MTFC on SAR image (courtesy of TESAR Program).

The effect of MTFC is not known. Figure 12.4, however, shows an SAR image before and after application of a sharpening filter. The filter has clearly added noise and degraded the image.

Equations (12.1) and (12.2) apply to a single-channel SAR operating at a single polarization. The equations also assume optimized processing, no bandwidth compression, and no artifacts due to sampling. As was the case for the GIQE, additional terms or adjustments would be required if these assumptions were not met.

12.5 Summary

The development of S&R modeling and performance prediction is far from complete. In fact, we are further behind now than almost anytime in history because of the rapid advances in sensor technology including SAR, IR, MSI, and HSI. Sensor development has far outpaced modeling activities. This leads to a situation where we are building systems without a clear understanding of cost/benefit trades.

Perhaps because of both its literal nature and its similarity to photography, with a much longer history of research, EO modeling for S&R appears most advanced, with IR not far behind. The situation is reversed for TAS with IR more advanced than EO. SAR modeling, in the context of predicting human observer performance is the least advanced. MSI modeling cannot be completed until the value of spectral data to the human observer has been more fully established. The GIQE, with appropriate calibration, can handle MSI through SWIR to predict MS IIRS ratings. The absolute benefits of other multidata systems (frequency, polarization, etc.) have yet to be defined.

The GIQE and static NVESD/PD models currently predict observer performance for EO and IR systems with an assumption that the target has been fixated and sufficient time is available for a recognition decision. The NVESD Acquire model adds a search term that has performed with varying degrees of accuracy. The GIQE does not account for the search process and no successful model of this process appears to exist. The nondeterministic nature of search makes it a difficult problem.

GIQE and NVESD/PD models also do not fully account for sampling and bandwidth compression effects, although other studies provide a basis for estimating the effects. Physique offers some potential in this regard, but validation data is not available. In the case of SAR, only the RTQF has been designed to predict observer performance and only for coarse resolution systems.

There is clearly a need for an SAR model as long as human observers are to be used. Although ATR systems may be desirable for SAR, their success has yet to be demonstrated. A conceptual model to predict radar NIIRS was presented; however, extensive development and validation are required.

Using the GIQE as a basis, a conceptual approach to the modeling of HSI has also been developed and validation is beginning [10,11]. Here, the human will probably not be involved in the information extraction process—this may actually simplify modeling development.

Although the temporal domain has utility in S&R applications, it has not been explicitly treated in S&R modeling. It appears that in most cases, one need only model the spatial aspects of the problem. The temporal aspect simply requires appropriate sampling. Thus, if the objects of interest can be seen and are imaged with sufficient frequency, the benefits of the temporal domain can be realized.

In the case of image-based models, both IQM and the Sarnoff JND models have been shown to correlate with visible NIIRS. Further development of both models would be required for IR and radar imagery, and more extensive validation of both would be desirable. It is not clear that the IQM could be extended to MSI; the JND model currently accounts for color with video, but with unknown success. Finally, none of the existing image-based models has been shown to predict subjective video quality ratings significantly better than the PSNR metric. This may be of little consequence in predicting the performance of S&R systems but may be of concern in understanding subjective image quality.

It is clear that many challenges remain in accurately modeling and predicting performance for S&R systems. Just as S&R systems will continue to evolve and potentially provide more information, the information gathering process will become increasingly difficult. The proliferation of weapons of mass destruction and terrorist activities have vastly complicated the S&R process. Innovative techniques will be required to meet these threats, and performance prediction and modeling advances will be necessary to assist in combating these national security issues.

References

[1] Farrell, R. J., and J. M. Booth, *Design Handbook for Imagery Interpretation Equipment*, Seattle, WA: Boeing Aerospace Company, 1984.

[2] Imagery Analysis Division, *Multispectral NIIRS Development*, Washington, DC: National Exploitation Laboratory, 1995

[3] Oliver, C., and S. Queqan, *Understanding Synthetic Aperture Radar Images*, Norwood, MA: Artech House, 1998.

[4] Waldman, G., J. Wooton, and G. Hobson, "Visual Detection with Search: An Empirical Model," *IEEE Transactions on Systems, Man, and Cybernetics*, Vol. 21, No. 3, 1991, pp. 596–606.

[5] Toet, A., P. Bijl, and M. Valeton, "Test of Three Visual Search and Detection Models," in W. R. Watkins, D. Clement, and W. R. Reynolds (eds.), *Proc. SPIE Vol. 3699, Targets and Backgrounds: Characterization and Representation*, 1999, pp. 323–334

[6] Enoch, J. M., and G. A. Fry, *Visual Search of a Complex Display: A Summary Report*, Colombus, OH: Ohio State University Research Foundation, 1958.

[7] Biederman, I., "Recognition by Components: A Theory of Human Image Understanding," *Psychological Review*, Vol. 94, No. 2, pp. 115–147.

[8]   Leachtenauer, J. C., " Peripheral Acuity and Photointerpretation Performance," *Human Factors,* Vol. 20, No. 5, 1978, pp. 537–551.

[9]   Pastore, R. E., and C. J. Schrirer, "Signal Detection Theory: Considerations for General Applications," *Psychological Bulletin,* Vol. 81, No. 12, 1974.

[10] Martin, L., J. Vrabel, and J. Leachtenauer, "Metrics for Assessment of Hyperspectral Image Quality and Utility," *ISSSR Proceedings,* Las Vegas, Nevada, 1999.

[11] Martin, L., et al., "Image Quality Evaluation of AOTF and Grating Based Hyperspectral Sensors," *ISSSR Proceedings,* Las Vegas, Nevada, 1999.

List of Acronyms

ABS	average boundary strength
ACE	average co-ocurrence error
ACTD	advanced concept technology demonstration
AFV	armored fighting vehicle
AMLCD	active matrix liquid crystal displays
AMRL	Air Force Medical Research Laboratory (USAF)
ANSI	American National Standards Institute
APC	armored personnel carrier
AR	artifact rating
BeSRL	Behavioral Science Research Laboratory (U.S. Army)
BWC	bandwidth compression
CCD	charge coupled devices
CC&D	camouflage, concealment, and deception
CD	compact disc
CDL	common data link
CNR	clutter-to-noise ratio
CODEC	code/decode algorithm or device
CRT	cathode ray tube
CSF	contrast sensitivity function
CTF	contrast transfer function
DARO	Defense Airborne Reconnaissance Office
DARPA	Defense Advanced Research Projects Agency
DAS	detector angular subtense
DCT	discrete cosine transform
D&D	deception and denial

DICOM	Digital Imaging and Communications in Medicine
DoD	Department of Defense (U.S.)
DPCM	delta pulse code modulation
DRA	dynamic range adjustment
DSCQS	Double Stimulus Continuous Quality Scale
DSIS	Double Stimulus Impairment Scale
DVQ	digital video quality (metric)
DWT	discrete wavelet transform
EGSD	equivalent ground-sampled distance
ENS	engineering NIIRS standard
EO	electro-optical
ERIM	Environmental Research Institute of Michigan
ERTS	Earth Resources Technology Satellite
FASCODE	Fast Atmospheric Signature Code
FED	field emissive display
FFT	fast Fourier transform
FIR	finite impulse response
FLIR	forward-looking infrared
FOR	field of regard
FOV	field of view
FPA	focal plane array
FROC	free receiver operating characteristic
FWHM	full width at half maximum
GIQE	general image quality equation
GPS	Global Positioning System
GR	ground resolution
GRD	ground-resolved distance
GSD	ground-sampled distance
HC	hardcopy
HSI	hyperspectral imagery
HRC	hypothetical reference circuit
HS	hyperspectral
HVS	human visual system
IA	imagery analyst
ICS	integrated contrast sensitivity
ID	information density
IDEX	Image Data and Exploitation System (or Facility)
IFOV	instantaneous field of view
IFSAR	interferometric synthetic aperture radar
INS	Inertial Navigation System
IPR	impulse response

IQM	image quality measure
IR	infrared
IRARS	Imagery Resolution and Reporting Standards (Committee)
ISAR	inverse synthetic aperture radar
ITU	International Telecommunication Union
JND	just-noticeable difference
JPL	Jet Propulsion Laboratory
JPEG	Joint Photographic Experts Group
LCD	liquid crystal device
LED	light emitting diode
LOWTRAN	Low Resolution Transfer Code
LSI	linear shift invariant
LWIR	longwave infrared
MODTRAN	Moderate Resolution Transfer Code
MPEG	Motion Pictures Experts Group
MRC	minimum resolvable contrast
MRT	minimum resolvable temperature
MS	multispectral
MSI	multispectral imagery
MTF	modulation transfer function
MTFA	modulation transfer function area
MTFC	modulation transfer function compensation
MTI	moving target indicator
MWIR	midwave infrared
NASA	National Aeronautics and Space Agency
NATO	North Atlantic Treaty Organization
NC	noise criterion
NEI	noise-equivalent input
NEMA	National Electrical Manufacturers Association
NEP	noise equivalent power
NETD	noise-equivalent temperature difference
NIDL	National Information Display Laboratory
NIIRS	National Imagery Interpretability Rating Scale
NIMA	National Imagery and Mapping Agency
NIR	near infrared
NSR	noise-to-signal ratio
NTSC	National Television Systems Committee
NVESD	Night Vision and Electronic Sensors Directorate
NVL	Night Vision Laboratory (Now Night Vision and Electronic Sensors Directorate)
PAL	Phase Alternate Line

PI	photointerpreter
PR	probability of recognition
PSF	point spread function
PSNR	peak signal-to-noise ratio
RABS	ratio average boundary strength
RADC	Rome Air Development Center (USAF)
RCS	radar cross-section
RER	relative edge response
RMSE	root-mean-square error
ROC	receiver operating characteristic
ROIC	readout integrated circuit
RTQF	radar threshold quality factor
S&R	surveillance and reconnaissance
SAR	synthetic aperture radar
SC	softcopy
SCR	signal-to-clutter ratio
SLR	side-looking radar
SMPTE	Society of Motion Picture and Television Engineers
SNR	signal-to-noise ratio
SPG	self-propelled gun
SQF	subjective quality factor
SQRI	square root integral (model)
SQS	subjective quality scale
SSM	surface-to-surface missile
SWIR	short-wave infrared
TAS	target acquisition sensor/system
TBR	target-to-background ratio
TDI	time delay integration
TESAR	Tactical Endurance Synthetic Aperture Radar
TQF	threshold quality factor
TSC	task satisfaction confidence
TSD	theory of signal detection
TTC	tonal transfer compensation
TTPF	target transfer probability function
TV	television
UAV	unmanned aerial vehicle
USAF	United States Air Force
USGS	U.S. Geological Society
USSR	Union of Soviet Socialist Republics
UV	ultraviolet

VCR	videocassette recorder
VEM	visual edge matching
VIE	visual image evaluation
VISDET	Visual Detection (search model)
VQEG	Video Quality Experts Group
VSM	video systems matrix
WAS	wide area search

About the Authors

Jon C. Leachtenauer received his A.B. and M.S. degrees in geology from Syracuse University. After serving as a photointerpreter in the U.S. Army, he began a 40-year career in human factors research in image quality and photointerpretation performance measurement. He has worked for Aero Service Corporation, Photics Research Corporation, and the Boeing Company. At ERIM from 1978 to 1999, he conducted research in SAR image quality and performance metric development, and for several years led the Sensor Research Department. While at ERIM, he was the Senior Scientist for the Imagery Analysis Division of the National Exploitation Laboratory. He led the development of the Radar NIIRS, established and led a display quality research program, and conducted numerous image quality and utility studies. He formed a consulting company in 1999 (J/M Leachtenauer Associates) and is currently a consultant to the National Imagery and Mapping Agency. He is the author of over 150 technical reports as well as numerous published papers covering all aspects of the image exploitation process.

Dr. Ronald G. Driggers has 12 years of electro-optics experience and has worked for or consulted with Lockheed Martin, SAIC, EOIR Measurements, Amtec Corporation, Joint Precision Strike Demonstration Project Office, and Redstone Technical Test Center. He is currently working for the U.S. Army's Night Vision and Electronic Sensors Directorate and is the U.S. representative to the NATO panel on advanced thermal imagery characterization. Dr. Driggers is the author of two books on infrared and electro-optics systems and has published over 30 refereed journal papers. He is co-editor of Marcel Dekker's *Encyclopedia of Optical Engineering* and is an associate editor of *Optical Engineering*.

Index

Recent Titles in the Artech House Optoelectronics Library

Optical Fiber Communication Systems, Leonid Kazovsky, Sergio Benedetto, and Alan Willner

Optical Fiber Sensors, Volume Four: Applications, Analysis, and Future Trends, John Dakin and Brian Culshaw, editors

Optical Fiber Sensors, Volume Three: Components and Subsystems, John Dakin and Brian Culshaw, editors

Optical Measurement Techniques and Applications, Pramod Rastogi

Optoelectronic Techniques for Microwave and Millimeter-Wave Engineering, William M. Robertson

Reliability and Degradation of III-V Optical Devices, Osamu Ueda

Smart Structures and Materials, Brian Culshaw

Surveillance and Reconnaissance Imaging Systems: Modeling and Performance Prediction, Jon C. Leachtenauer and Ronald G. Driggers

Tunable Laser Diodes, Markus-Christian Amann and Jens Buus

Understanding Optical Fiber Communications, Alan Rogers

Wavelength Division Multiple Access Optical Networks, Andrea Borella, Giovanni Cancellieri, and Franco Chiaraluce

For further information on these and other Artech House titles, including previously considered out-of-print books now available through our In-Print-Forever® (IPF®) program, contact:

Artech House	Artech House
685 Canton Street	46 Gillingham Street
Norwood, MA 02062	London SW1V 1AH UK
Phone: 781-769-9750	Phone: +44 (0)171-973-8077
Fax: 781-769-6334	Fax: +44 (0)171-630-0166
e-mail: artech@artechhouse.com	e-mail: artech-uk@artechhouse.com

Find us on the World Wide Web at:
www.artechhouse.com